Wild Plants of Eastern Canada

Marilyn Walker, PhD

NIMBUS PUBLISHING LTD

Nimbus Publishing Limited
PO Box 9166, Halifax, NS B3K 5M8
(902) 455-4286 www.nimbus.ca

Printed and bound in Canada
Nimbus Publishing is committed to protecting our natural environment. As part of our efforts, this book is printed on 100% recycled content, Ancient-Forest Friendly, Enviro 100.

Design: Kathy Kaulbach, Touchstone Design House
Author photo: Wayne Poulsen

Library and Archives Canada Cataloguing in Publication
Walker, Marilyn, 1949-
Wild plants of Eastern Canada : identifying, harvesting & using / Marilyn Walker.
Includes bibliographical references and index.
ISBN 978-1-55109-615-5

1. Wild plants, Edible—Canada, Eastern. 2. Wild plants, Edible—Canada, Eastern—Identification. I. Title.

QK203.E37W34 2008 581.6'3209713 C2008-900013-7

Canada

The Canada Council for the Arts | Le Conseil des Arts du Canada

NOVA SCOTIA
Tourism, Culture and Heritage

We acknowledge the financial support of the Government of Canada through the Book Publishing Industry Development Program (BPIDP) and the Canada Council, and of the Province of Nova Scotia through the Department of Tourism, Culture and Heritage for our publishing activities.

Dedication

To Morgan and Taylor, in memory of their father, Peter Walker.
May you grow to love plants and may they sustain you and nourish you on your life's journeys, as they have mine.

Table of Contents

Preface:
How to Use this Book

This book, written for non-experts and amateur naturalists as well as for experienced gardeners, covers some of the more readily available edible and medicinal plants of northeastern Canada and the United States. (The book should be useful even if your wilderness rambles aren't within the region, as many of the plants included here are found in other parts of North America, and even in other parts of the world.) Some of the species are native to the region, others have been introduced, and still others are escapees from gardens and have now become naturalized. From a vast number of possibilities, I have chosen plants that for scientific, aesthetic, or practical reasons attracted my own interest since I made my first visit to the Maritimes in 1967.

This book also presents—in an accessible, readable way—some of the history associated with the plants I selected. I made several visits to Britain to research the origins of plants that are now commonplace here, visiting the archives of British museums and libraries and the medicinal gardens at Kew, Chelsea, Oxford, and Cambridge. I found references to the region's plants that had been "discovered" by newcomers and noted in explorers' journals and naturalists' diaries. Quotations are included from the writings of these explorers, missionaries, traders, adventurers, and other observers who took the time to record information about the plants they encountered as they explored what was, to Europeans, a new and unknown land. I also located actual plants, or their offspring, that had been pirated back to the Old World for their investigation and propagation.

Also included in this book is information I gathered first-hand from First Nations healers and elders whom I have been fortunate to work with over thirty-five years (some of whom have taught my university courses). I have also taken advantage of the many scientific studies that have been carried out in recent years on the medicinal properties of plants. Finally, I have blended my interests in nature-based methods of farming and gardening to suggest—in a spirit of cooperation and mutual understanding—ways for us to deal with some of the environmental problems facing the region.

The style of this book is non-technical. Because it is intended for the layperson rather than the botanist, certain liberties have been taken. The world "berry," for example, is used in the popular rather than the strictly botanical sense. Plants are listed in alphabetical order by their most common English names, rather than according to botanical classification. Many plants—marsh marigold, yellow pond lily, and shrubby cinquefoil, for example—

have adjectives that are an integral part of their names. In these cases, the plant is listed under the first part of the name (e.g. yellow pond lily is listed by "yellow"). So many plant names are prefaced by "wild," however, that in this single instance, the descriptive word is ignored in organizing the listings. Thus, wild rose is listed by "rose" and wild chives by "chives." The index lists all the plants for easy reference.

Latin names for each plant are also included in the field guide as an exact and universal means of identification. Latin names come in two parts. The Latin name for strawberry blite, for example, is *Chenopodium capitatum. Chenopodium* is the name of the genus (a genus represents from one to many species); *capitatum* is the species name. After the first use, a genus may be referred to by its first letter only: *C. capitatum.* Any letters or abbreviations following a species name indicate the person who most recently determined that species. Latin names are useful since a plant may have many common names in different parts of the country or at different times in history. The opposite can also be true: entirely different plants may be known by the same common name. For example, kinnikinnick may refer to both *Arctostaphylos uva-ursi* (bearberry) and *Cornus amomum* (silky dogwood), the latter a species that does not grow in this region. The use of a common name becomes further complicated when you consider that a common name may refer to a genus or to a particular species. For all these reasons, it's often best to refer to the Latin or scientific names that are consistent everywhere. For further clarification, plants in the field guide are also identified by the family (a group consisting of more than one genus) to which they belong.

The primary reference used in the preparation of this guide is *Roland's Flora of Nova Scotia* (revised by Marian Zinck in 1998). This two-volume guide does not include all northeastern plants, as it focuses on Nova Scotia, but it is an excellent (and weighty) botanical reference. While referencing other material is always a good idea, keep in mind that other sources may present conflicting information. Just because a reference is old doesn't mean it is accurate—knowledge changes over time, as do plants and their habitats. The Northeast has had longer and more intense European influence than elsewhere in Canada or the U.S. It is, and always has been, a dynamic environment. Plants (like people) are always responding to new and changing conditions, and their distribution is constantly being affected by movements of people. Traditional plant habitats may have been destroyed or damaged by logging, farming, or industry, or they may have reacted to altered soil and climatic conditions.

Even today, the properties of many plants are not well known, and the range of a particular species may not be fully documented. As a result, much more information is available for some plants than for others. Gaps in our knowledge are filled in with every botanical expedition we make, and old knowledge is constantly being updated. No matter how many times I go out to gather mullein, or blueberry, or Labrador tea, I learn something new every time. Our primary source of knowledge is the plants themselves.

The Northeast is still a land to discover and explore. I hope this book will awaken every reader's sense of adventure. May it serve as a reminder of the richness of our botanical past and provide us with guidelines for conserving our future.

The Illustrations

In this book, plants are illustrated using the traditional technique of leaf-printing, sometimes called nature-printing. The origins of this technique are perhaps lost to us, but may go back as far as the hands that were pressed first in pigment and then onto the wall of a cave. Plant printing for the purpose of book illustration goes back as far as the early 1200s in Europe, with Leonardo da Vinci being the most well-known scientist associated with this technique. I chose leaf-printing for this book because no other medium gives such detail. I hope that when you look carefully at the prints of the leaves, you will be as awestruck as I was at the intrinsic beauty of nature.

Acknowledgements

Many people shared their expertise and experiences with me over the years. They cannot all be mentioned here by name, but I am grateful for their teachings. The following I would like to thank individually:

Mildred Milyea
Jeorgina Larocque
Blair Léger
Michelle and Claude Parent-Gagné
Émile and Evalyn Gautreau
Charles and Judith Hubbard
Gordon LaBillois
Margaret LaBillois
Fred Isaac
Skip Koolage
Earl LaBillois
Ann LaBillois and Mary Rebecca LaBillois
Marion LaBillois
Nora LaBillois
Stephanie LaBillois and her mother, Shirley Bear
Chief Mary Lou Parker
Bob Sam
Suzanne Schroder
Joe John and Irene Sanipass
Bernard Francis
Mister Sleep
Mother Earth Lodge
Bill Lewis
Chief Frank Meuse
Ray Solomon
Bob and Norma Frelick
David Smith and Save a Plant
Ed Lumgair
Ruth Miller
Clara Clayton-Gaugh, who was interviewed for a class project by Beth Llewellyn.
The many students I explored with in my ethnobotany classes.

Very special thanks to Elaine Moolenburgh and Wayne Poulsen.

Introduction:
The Forgotten Values of Wild Plants

As we become more and more aware of the benefits of eating locally and of growing our foods and plant medicines without artificial pesticides and fertilizers, we are beginning to recognize the value of the wild plants that surround us. We have taken them for granted, but they have much to teach us about working with nature rather than against it. Plants have become important in new models of health and wellness—being in nature can be as beneficial to our health as taking a medicinal tea or applying a salve. We are also beginning to value the ecology of plants and plant communities. Plants are useful not only to humans, but also to fish, birds, insects, and other animals for food and shelter. Plant succession and ecological restoration is also becoming more commonly understood, though we still have a great deal to learn from the plants themselves about nature-based and sustainable methods of plant propagation and care.

Raspberry

The benefits of harvesting wild plants are many. In addition to providing sustenance and variety, wild plants can stretch a limited budget. Raspberry leaf tea, for example, is an expensive item in specialty shops, yet it can be made from leaves picked free of charge just about anywhere in Canada. In the Maritimes, much of our produce is imported from as far away as California and Mexico. As well as being bountiful, local wild greens are much higher in nutritive value than produce that has lost its freshness and nutrition from cold storage, transportation, processing, and the addition of herbicides and pesticides to the soil. Although information on wild foods in terms of their vitamin, mineral, protein, carbohydrate, and fibre content is sometimes incomplete, we know that many wild food plants are nutritionally equal or superior to cultivated plants. Examples abound: dandelion greens and lamb's quarters are higher in vitamin A than spinach; rosehips are one of the best sources of vitamin C in the world. Also, many wild greens are ready for harvesting a month or more before garden crops are ready in the spring—a real advantage for colder areas of the region. The ability of native plants to withstand frosts that cultivated plants cannot survive also means that they will flourish late into the fall when gardens are finished for the year. And many wild fruits and vegetables can be frozen, dried, canned, or made into preserves for year-round use, just like homegrown produce.

Some wild plants are easily cultivated if you want them accessible just outside your door. Wild chives transplant easily into a garden or planter box. Weeds such as lamb's quarters, dandelion, and chickweed will grow in waste areas or on stony ground where other plants can't take hold. They won't need watering or fertilizing, and they will even prepare the soils for garden plants to thrive (some even restore damaged areas). And perhaps most importantly, there's none of the work associated with cultivating your own fruits and vegetables!

Plants as Medicines

Traditional knowledge about using plants as medicines is detailed and complex. It is based on the changing qualities of plants from season to season, over time, and within different habitats. While both traditional medicine and the use of nature-based remedies are growing in popularity, non-Native people still have a lot to learn about plant medicine from Aboriginal Peoples. At the same time, we must be careful to recognize intellectual property rights in appropriating such knowledge and using it out of context, particularly because, by tradition, it is passed on orally. We also must consider how the excessive use of medicinal (or any other) plants poses threats to biodiversity locally and globally. Traditional medicine recognizes that plant parts work in combination, whereas western medicine has been primarily interested in extracting a single component of a plant. Global standards are being developed about the regulation, administration, and protection of medicinal plants and their place in Indigenous People's cosmology. Researchers are also assessing how this knowledge may be integrated into western medicine for the benefit of both indigenous and non-indigenous peoples.

I've learned many things from my contacts with First Nations and Native Americans over the years, none more important than the protocol of respect—asking a plant if it may be used and giving a few grains of corn meal or rice as a sign of respect. (Traditionally, native tobacco is used as an offering.) I have always been told to empty my heart of bad thoughts before I go collecting plants for medicine, and to smudge myself with the smoke of juniper, sweetgrass, or wild sage as a way of purifying and preparing myself. The most enduring lesson has been that plants are gifts from nature, from Mother Earth.

A variety of wild sage

Some Rules for Harvesting Wild Plants

Gather only healthy-looking specimens for food or medicine. Stay away from those that are blemished or wilted. Make it a rule not to collect in sprayed areas, dusty roadsides, and garbage dumps. Aquatic plants should be growing in clean water, far from sewage, dumping grounds, and docks. If, in an emergency, you are uncertain of the purity of the water, boil both water and plants, or purify the washing water with a few drops of chlorine bleach or iodine. Store what you have collected in a cool place until it's ready to use. Some berries (such as raspberries, strawberries, and cloudberries) will last longer and taste better if you don't rinse them. But if you need to, wash them just before using, and lay them out on a clean cotton towel to absorb the excess moisture. If anything develops mould, discard it.

Finally, be sure you know what you are gathering. If you are interested in edible wild plants, you should be interested in the poisonous or potentially toxic ones too. Sometimes, the properties of a plant that make it "medicinal" in one dose can make it toxic in another.

In our enthusiasm for harvesting wild plants, let's remember that the supply is exhaustible and that plants vary in their ability to withstand harvesting. Care must always be taken to ensure the variety and continuity of plant (and animal) life in the Northeast. Steering clear of weakened plants will ensure that you benefit from your work but it also means that the plant will have a chance to recover from its ailment. When I've collected with First Nations, they've told me to be respectful of a plant that has already "given" of itself for food or medicine. A plant that is weak from insect damage, exposed roots, or pollution will need all its strength for itself. Even common, hardy plants should be harvested in ways that do not endanger their survival. A good rule of thumb is the Rule of Seven that was taught to me by First Nations: Never gather more than one out of seven plants in any area; if there are less than seven, look for another area. This ensures enough of a species is left for reproduction for the next seven generations, First Nations say. Always leave the "mother plant" untouched—as a sign of respect and so that it may continue to produce offspring. You can usually recognize the mother plant as the sturdy, mature plant that will be surrounded by her offspring.

Remember that collecting berries affects a shrub's ability to reproduce, even though the plant itself will not be killed. Taking the flowers may weaken a plant and reduce the production of seeds (which develop from a flower head) and thus affect the plant's long-term survival. Digging up a plant for its root will destroy the whole plant, so select your specimens with care. For some plants you can harvest a section of the root and leave the rest to continue to grow. If you're harvesting leaves, ensure some are left on the stem so the plant can manufacture the food that gets it through the winter.

Remember, too, that wildlife relies on the plants for survival, whereas you—under most circumstances—will be collecting for pleasure. Caribou rely heavily on lichens for food; when a heavy or ice-encrusted snow cover prevents caribou from reaching their usual food source, tree lichens are especially important to their survival. Pond lilies are food for moose in summer and for beaver and muskrat all year round. In winter, moose live primarily on willow, but may browse on young, tender white spruce seedlings. Berries are an important source of food to birds and mammals throughout the year. Even so-

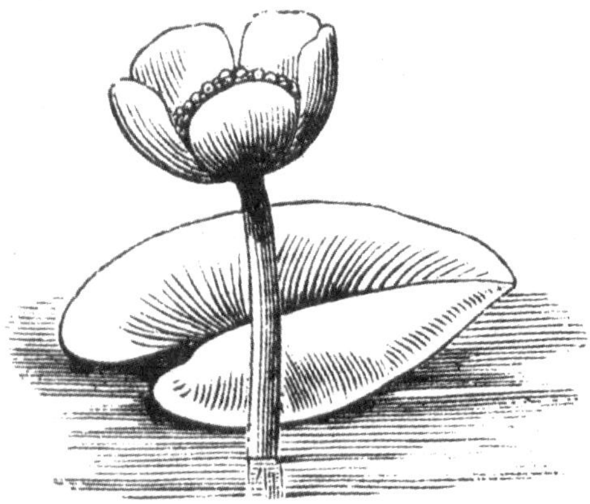

Yellow pond lily

called "weeds" provide food and shelter to mammals, birds, and insects. Consider the diet of the ptarmigan. In winter, it feeds on buds and twigs of willow and other small shrubs. In summer, it eats seeds and insects, and has been known to feed on the flowers of blueberry shrubs. Ptarmigan are an important link in the food chain in Newfoundland and Labrador, converting plant protein into animal protein—food for carnivores and predatory birds as well as humans.

Many of the plants now firmly established in the Northeast are recent introductions to North America. Some arrived from overseas as seeds in packing materials, in feed for livestock, or in bags of seed intended for planting. Dandelion and lamb's quarters are among the hitchhikers that spread quickly across the continents from this region. Others, for example clover, were imported specifically for cultivation in farmers' fields and home gardens, but soon "escaped" and are now "naturalized." Weeds are prime foods for wild plant foragers as they are widely available, plentiful, and easy to find. They can withstand more intense harvesting than many native species, which have more specialized habitats and cannot tolerate disturbance or intrusion by the more aggressive invaders. Even these common plants, however, should be harvested in a way that does not endanger their survival.

Transplanting from the Wild

Many plants that were once indigenous to the region are becoming rare or restricted in their locales. Intensive agriculture, recreational use of wilderness areas, over-picking, and heavy cutting of the forest have all harmed native plants and their habitats. Transplants can be a living laboratory for us to learn more about a particular plant—its likes and dislikes, how it responds to care, and whether it can be a future transplant back into the wild.

Plants should not be dug up from the wild, however, unless the circumstances are unusual, a very large concentration of the plant exists, and the permission of the property owner has been granted. A notable exception would be "salvage ethnobotany" that may be appropriate where a proposed construction site will destroy the plants. Otherwise, buy native species for transplanting from a reputable nursery that does not wild-gather, or propagate them yourself by following some simple guidelines.

If you identify a plant in the wild from which you plan to take some seed, mark it (perhaps with a ribbon around its stem) so that it can be located when the weather changes. Gather the seeds when the capsules or seed pods are dried out but before they split open and scatter, and before the frosts get heavy, which will make the plants go mushy. Choose a dry day and if the seeds are in a sunny spot, wait until the afternoon—the longer they are exposed to the sun, the drier they will be.

Place a large paper bag over the stem, scrunch it up around the seeds, and shake it until you hear the seeds drop into the bag. This works well for something like mullein or evening primrose. For other plants, you can snip the pods off with pruning shears.

Label the outside of the bag with a magic marker and put the bag in a dry, out-of-the-way place—on top of a cupboard or on a table near (but not too near!) a wood stove. On top of the fridge is a good spot, as the bag will get a mild dose of heat, which will help to dry the seeds further. After a few weeks, tip the contents of each bag onto a sheet of newspaper and sort through the contents to separate the seeds from the chaff. You'll find all sorts of ways to do this. Most seeds are heavier than the chaff so you might be able to blow the debris away with your breath. Or form a cone from a sheet of newspaper or from the original bag, cut a small hole in the base slightly bigger than the seed, and let the seeds drop into a container.

I keep my seeds in small paper bags or recycled envelopes so that moisture can still evaporate; plastic can cause mould if the seeds are not completely dry. For dry seeds, I find plastic film canisters to be useful for storage. You can write on them with an indelible marker. To further absorb moisture, add a few grains of powdered milk to each container.

Some seeds can be scattered directly into your garden in the fall; others should be kept until the spring. Store them in the refrigerator or even in the freezer to mimic the chilling they would get outdoors in their natural environment. They can be seeded into flats until they're ready for transplanting.

Conservation Reminders

1. Never collect more than you need.
2. If you find only a few plants (less than seven is a good rule of thumb) in one area, leave them and search elsewhere.
3. Do not harvest all the plants from one spot; always leave some behind.
4. Harvest only that part of the plant you need. If you're gathering the leaves, don't disturb the root.
5. Be careful not to disturb the surrounding area. Push branches aside instead of breaking them. Tread lightly, and look around you before you sit or put down your pack.

Part One:
Plants and People of the Northeast

Like Plants, Like People

Like people, plants are great colonizers. They wander about, singly or in "herds," trying out new conditions until they find somewhere to settle in. If the environment is not suitable, they go elsewhere out of necessity, or perhaps out of the same sense of adventure that inspired human migration. People have been using plants for as long as we've been people. Before we came along, animals also co-evolved with plants, relying on them for the essentials of life—food, medicine, and shelter. And before animals, insects and plants developed an evolutionary mutualism to ensure their own, and each other's, survival. It's fair to say that many plant species could no longer survive without human intervention, at least in the regions they have now spread to. Some plants, after all, can only be grown in greenhouses today.

From the seed ferns that first colonized the land millions of years ago to the ornamentals we plant in our gardens today, plants have adapted to the changing conditions that characterize the history—ancient and recent—of a given region. The landscape we now know as the Northeast is the end result of a complex interweaving of natural and human forces. And it is still evolving. A plant's ability to become established and to survive is affected by geology, natural forces such as fire, floods, gales, and snows, and more recently, humans.

One of the ways early European visitors inadvertently affected plant life in the Northeast was with the introduction of exotic animals to the region. Reports from Sir Humphrey Gilbert's visit to Newfoundland in 1583 describes the deposition of cattle by the Portuguese thirty years earlier. Sable Island became home to animals—pigs and horses among them—from shipwrecks. By the early nineteenth century, rabbits reached Sable Island and multiplied quickly, as did rats. Cats were brought in to eradicate the rats, and then dogs were brought in to deal with the cats! All of these animals impacted the

Dandelion

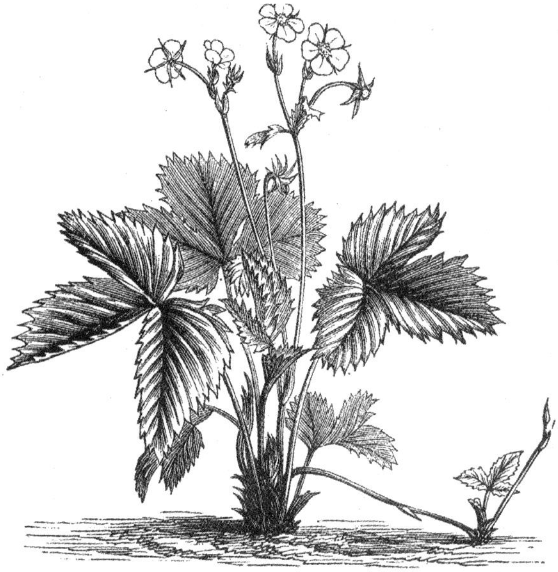

Wild strawberry

plant life in the region irreversibly, sometimes destroying the native species entirely.

Other plants have thrived with human activity. Most notorious "weeds" are nothing more than plants that have moved into our gardens and lawns where they are not wanted. There is no botanical category for them but they do share certain characteristics. They are very good colonizers, usually produce large numbers of seeds, and propagate under severe and limited conditions. Like all plants, they are great teachers if we approach them with respect and gratitude. Dandelion, for example, has a long taproot that, if broken by a digging implement, propagates from the pieces. While we do everything we can to rid our lands of it, we would do much better to work with it. Its role in the landscape is restorative, since its long root can penetrate into the deepest layers of soil and bring nutrients to the surface. It does well, as do many weeds, in damaged ecosystems where the nutrients in the top layers have been washed away through excessive lawn cutting or the use of pesticides and fertilizers.

Many seaweeds are just as important for their colonizing abilities, especially where they stabilize areas vulnerable to erosion. The next time you go to a beach or an island, look for senecio on the shore. Senecio's stout, fibrous roots anchor the plant in the shifting sands, and also stabilize the dunes. Sea plantain, with its tough, spiky leaves, is another beach plant common to the region that helps to stabilize the soils. Eel grass is a curious aquatic plant that grows between sea and land, trapping silt and bits of detritus in its roots to build up its own habitat. Sea plants in general are avid colonizers—the currents and tides carry their spores and distribute them evenly and widely. Irish moss, for example, densely occupies the entire coastline of the Atlantic shore, and many of our other seaweeds have a worldwide distribution.

There are certain other plants that are always the first to colonize an area. Fireweed is so-named because it is one of the first plants to colonize after a fire. Wild strawberry's little rootlets, from which new plants are produced via runners, can take hold in very little soil, and often get a hold in small cracks in concrete or between paving stones. Wild raspberry will grow very tall (up to six or seven feet high) for the first four or five years after a forest fire or insect infestation. These plants gradually grow back to normal size as trees begin to make a comeback ten years or so later. All three plants are important as the vanguard of plant succession; they build up the soil base for subsequent species.

The Northeast as a Region

It is difficult to give definitive geographical or cultural boundaries to the region covered in this book. Some researchers say its northern extent is the southern margin of the boreal forest that extends in a wide band across Canada and Alaska. That leaves out the Innu peoples of Labrador and Labrador itself, however, and ignores those parts of the Atlantic provinces that are home to coniferous forest. The region also includes a maritime ecology quite different from that of the boreal forest. The southern boundary of the Northeast as a region is even more arbitrary and is complicated by the overlay of the United States/Canadian border, which does not take into account Aboriginal territories. Rather than trying to give a common definition, I try in this book to give a sense of the cultural, ecological, and historical diversity of a region that, to the people who live here, feels like home.

This is why you'll find terms like the Maritimes, the Atlantic region, the eastern seaboard, the northeastern United States, southeastern Canada, and so on, all used in different sources to refer to the same, or not quite the same, region. I consider the area to be the traditional homelands of the Beothuk, Mi'kmaq, Maliseet, and Passamaquoddy (often named as Maliseet-Passamaquoddy) First Nations. You'll also find references in which Aboriginal Peoples (First Nations in Canada; Native Americans in the United States) are "allocated" by anthropologists to a different "culture area"—as with the Innu and Inuit of Labrador. Let us remember, too, that this region has changed considerably through time.

The Atlantic shore of the Maritimes is defined by natural boundaries—to the north, the Quebec shoreline; to the east, the fence of the Labrador current, which extends down the east coast of Newfoundland; to the south, the barrier of the Gulf Stream; and to the west, the Appalachian land mass starting in Alabama and extending up the eastern seaboard through Newfoundland. It is unified as a region by the presence of the sea; at the same time, it is marvelously diverse. Severe gales and daily temperature extremes bring a notorious unpredictability and severity to the weather: rain falls every third day on Prince Edward Island, New Brunswick's summer highs might reach one hundred degrees Fahrenheit, and Newfoundland has an annual snowfall of about ten feet.

The northeastern continental shelf (stretching from Newfoundland to southeast Nova Scotia) houses many underwater plateaus, or banks, with shallow waters. The largest of these plateaus are Georges Bank in the Gulf of Maine and the Grand Banks of Newfoundland (one of the widest of all continental shelves anywhere in the world). The Grand Banks are a sort of fulcrum on which human settlement has been built. Here, the cool south-flowing Labrador current coming down from the Arctic blends with the warm Gulf Stream from the temperate south. The high mineral content infusing from the ocean's floor provides excellent conditions for plankton to proliferate; the plankton in turn support an abundance of marine life and the fish that rely on it. The excellent fishing on the Grand Banks provided Aboriginal Peoples with a rich and reliable resource and drew Europeans to the region as far back as the Vikings.

Shore and sea are intricately balanced in the Northeast. Fresh water mixes with salt. Birds of shore and sea gather together to nest. Fish travel upstream from the oceans to spawn; in some places landlocked salmon, char, and smelt can be found inland. Many land plants are adapted to salty conditions. The salt hay marsh, for example, is nutrient-rich from the micro-organisms carried in on the tides. The tides bring in nutrients from the sea floor and deposit tiny animal life such as shrimp and saltwater worms in the creeks that fill at high tide. These are food for fish and shore birds. In turn, the mud flats and salt marsh ecosystems produce oxygen-rich algae, which are carried outwards by the tides to feed fish, birds, and a variety of bottom feeders. The tidal flow creates the tidal marshlands, which once covered thousands of acres before the Acadians began, in the seventeenth century, to dyke and drain. The Bay of Fundy, which extends into Nova Scotia and New Brunswick, acts as a funnel, compressing tidal action and producing the highest tides in the world. They are so powerful that as early as 1607, Samuel de Champlain's compatriots near Annapolis Royal were able to build what was probably North America's first mill, most likely for grinding grain.

The Paleoenvironment • Rock formations found in the Northeast represent most of the earth's history. In parts of the region, rocks were laid down in the Cambrian period 540 million years ago (mya), more than 100 million years before plants colonized the land.

The Carboniferous period (360 mya to 290 mya) produced the coal and natural gas deposits that the region's economy benefits from today. It also produced the fossil cliffs of Joggins, Nova Scotia, located near the head of the Bay of Fundy. The tides here constantly wear away the cliffs, continually exposing fossils at the surface. The cliffs are particularly famous for fossilized tree trunks found in their original positions. They also provide the first evidence that land animals (in this case, one of the world's first reptiles) lived during this period. The Joggins beach near the cliffs is a wondrous walk when the tide is low. You never know what you will find.

In the Mesozoic era, known as the "Age of the Dinosaurs" (251 mya to 65 mya), flowering plants first appeared as a food source for insects. Plants gradually diversified and flourished, evolving into the huge variety we find today in the region.

The Pleistocene epoch (3 mya until about 10,000 years ago) was the most drastic of the geological ages, as it compressed massive changes into a short time frame of about one million years. At the beginning of the Pleistocene, the Atlantic climate began to shift—the winters became longer and colder. Eventually the region was covered year-round with snow. Under its own weight, this snow turned to ice and began to move down hills and into valleys, carrying huge deposits of soil with it. Eventually, the deep inlets it created became sanctuaries for human habitation. People lived and hunted at the entrances to these inlets in the summers, moving to their more sheltered interiors in the winters. Humans had arrived!

The melting ice left dry land that was colonized by plants and animals, although this took many years in some areas and even then, the effects were uneven. In some areas it took hundreds of years for trees to take hold; in Newfoundland, it took thousands of years. In some parts of the region, the deposition of glacial matter interacting with the rocks beneath created areas of low fertility and poor drainage that could support only conifers and some hardwoods. With slight alterations in the earth's landforms (such as geomorphic tipping), improved drainage meant that yellow birch, sugar maple, and beech could thrive.

The Pleistocene produced mineral wealth, but conditions for agriculture were poor except in a few areas. Generally, Maritime soils are "podzols"—ash-like soils that result from short summers, cold winters, and heavy rains interacting with the forest cover. As water percolates, it removes soluble substances from the surface and washes them away. The cold limits the decaying activities of micro-organisms and so plant matter remains on the surface as raw humus. Podzolic soils, even those in Nova Scotia's agricultural centre of the Annapolis Valley, are fairly low in fertility and can sustain only a small variety of crops, among them apples.

The "barrens" concentrated in Newfoundland, Labrador, and Nova Scotia are the result of several other processes interacting with the peculiarities of the acidic, coarse soils, rocky ground, and remaining permafrost from glaciation (processes including the repeated burning by early settlers that killed off tree growth). In my view, however, the term "barrens" is a misnomer. I find these lands to be rich and fascinating ecosystems, and they are home to plant communities not found elsewhere. Their "tiny trees"—dwarf alder, birch, and willow—grow no taller than ground-cover height, and berries proliferate. One of my happiest plant-collecting days was spent at Blueberry Point in New Brunswick, waist high in blueberry bushes. The berries, in fact, derive their medicinal properties from the particular qualities of the soil. Labrador tea, crowberry, laurel, sarsaparilla, and the curious pitcher plant are some of the barrens plants that Aboriginal Peoples benefited from long ago and still use today.

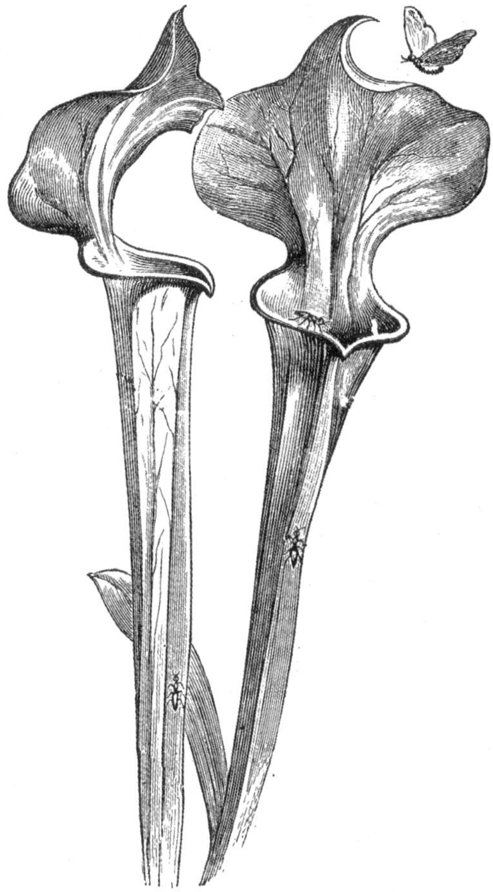

Pitcher plant

Human Habitation of the Northeast

The best-known archaeological site in the Maritimes is at Debert, Nova Scotia. To the Mi'kmaq, the peoples who used this site are known as the Ancient Ones. Archaeologists term them Paleoindians and dated this earliest human habitation of eastern North America at between 10,800 and 10,050 years ago at the end of the Pleistocene. These peoples probably moved northwards as the climate warmed and the glaciers retreated (by this time, the glaciers may have been only miles away). Debert was probably a seasonal camp for intercepting woodland caribou on their annual fall migration, since the Paleoindians were caribou hunters, making stone tools from chalcedony.

Britons, Africans, and the Irish may have reached the shores of the North Atlantic

Wild grape

between 500 BC and AD 500, but the first confirmed European settlement was at L'Anse aux Meadows about AD 1000—probably Viking. Fishing brought the Basques into the Grand Banks prior to 1492 but most likely they had little contact with Aboriginal Peoples. In 1492, Christopher Columbus landed in the New World, erroneously calling the inhabitants he met "Indians" because he thought he had discovered a western passage to the Orient. After Columbus, European fur traders and fishermen increasingly dominated the area so that by at least 1500, English, Breton, Basque, Norman, and Portuguese were fishing off Newfoundland and Labrador and Nova Scotia. After 1520, Aboriginal-European relations intensified with the beginnings of the fur trade. The first long-standing European settlements were soon established in the northern Maritimes. Religious conversion, especially by the French, soon followed. By 1611, the Jesuit mission at Port Royal in Nova Scotia began its activities, helping ensure First Nations' allegiance to the French.

Contact between Europeans and Aboriginal Peoples had devastating consequences for the health of First Nations, but Europeans, too, suffered from epidemics and a lack of medicines to treat them. Between 1616 and 1619, up to ninety percent of the population of southern New England succumbed to disease, possibly the bubonic plague or yellow fever. The ongoing search for medicines led to many North American plants being taken back to Europe for investigation. London's Kew Gardens, the Chelsea Physic Garden, and the medicinal gardens at Oxford and Cambridge were all recipients of New World plants carried back from the Northeast.

But without question this exchange was small compared to that moving from east to west: new species of plants and animals were introduced to the New World from Europe by the score. Settlers brought seeds of their favourite plants from home to grow in their New World gardens. Some of these escaped and became naturalized to the point that they appear as if they've always grown here. Others were introduced unintentionally as a hitchhiker in the pockets of a garment or as an ingredient in the livestock feed that ships brought along.

The Vikings • The Vikings came upon a new land that was rich in timber, game, fish, and fields of wild wheat with well-established communities of Aboriginal Peoples. The Vikings named it Vinland, perhaps referring to wild grapevines. Wild riverbank grape (*Vitis riparia*) was found growing on sites such as L'Anse aux Meadows, although today, wild grape grows no farther north than southern

New Brunswick. Grapes are mentioned in Viking sagas, an oral record that was written down in the late 900s by Christian missionaries sent to convert the pagan Vikings. In the missionaries' view, the Norse were terrorizing heathens, their agricultural and pastoral achievements not worthy of acknowledgement. Consequently, we know more about their military, navigation, and trading exploits than about their plant knowledge.

The trees that grow in the region today—dwarf balsam, larch, fir, poplar, and birch—were much the same in the eleventh century. The peat or sphagnum bogs near which the Vikings located their buildings gave them building material for their Icelandic-style homes—layers of sod over a timber frame. The same structures were used as burial places for bodies and belongings. Birchbark was kept in rolls, perhaps to be used as a wrap for stones used as net sinkers. Ropes, and perhaps barrel hoops, were made from spruce roots. The Viking presence in North America did not last, however. Greenland and Europe apparently held greater interest.

Blacks • Black history in the region goes back to the beginnings of European settlement—Mathieu Da Costa was an interpreter of African descent who accompanied Champlain, but the contributions of blacks to the Canadian multicultural mosaic and to the region's ethnobotany have not been given the attention they deserve. During the French regime, blacks lived as slaves, first settling the region in 1604 with French explorers on the island of St. Croix (an unsuccessful settlement that moved to Port Royal). As slavery was then legal in France, it would not have been unexpected for the colonies to have black slaves; blacks may thus have been the first settlers in Acadia. After French dominance of the region subsided, blacks lived as slaves of American and English immigrants. During and following the American War of Independence, they lived in the Northeast as free citizens and Loyalists (British supporters during the American Revolution), and as refugees from Jamaica and the U.S. Today, Nova Scotia has the largest black population of the Maritime provinces, at about eleven percent of the total population (twenty-five to thirty thousand people).

Most of the blacks who lived in the region did not receive property, land grants, or provisions, and if they did, they received only a portion of that allocated for whites. The rich farmlands of the Annapolis Valley, for example, were reserved for whites, while Black Loyalists were settled in marginal areas where agriculture was difficult. Farming thus never became the means of advancement it was for whites. Blacks survived such adverse conditions by finding crops that could grow in poor soils and becoming involved in sharecropping, developing alternative ways to enter the market economy. In the Halifax markets, they sold wild berries (cranberries, blueberries, strawberries, and blackberries), greens, potatoes, turnips, carrots, and tomatoes. Crabapples grew naturally in some areas and orchards were also planted. Fruit was sold fresh and made into preserves.

Woodcutting was an important source of income for blacks, as was the manufacture of wooden tools and implements such as brooms, oar rafters, hoop poles, evergreen wreaths, and baskets. Basketry also became a specialized art, with blacks teaching First Nations how to make some of the beautiful woven baskets (which the First Nations then adapted to make their own) for which the region is now famous. Home remedies introduced by blacks in the region included mustard plaster, which was made from the seeds of the mustard plant mixed with flour and shortening to make a paste; it was applied as a compress to ease arthritis and muscle spasms.

Acadians • Champlain sailed into the Annapolis Basin, noting, "This place was the most suitable and pleasant for a settlement that we had seen." Thus was founded a settlement at Île Sainte Croix (now in the United States) in 1604, the first attempt at permanent settlement north of Florida. (It would move to Port Royal after a disastrous winter.) Salt marshes, with their twenty thousand hectares of fertile soils, were the basis for French colonization of the region. Mi'kmaq referred to the region as Chignecto, meaning "great marsh district." The Acadians, its first settlers, called the marshes "tintamarre" (meaning racket or din) because of the vast numbers of birds they supported.

Of the original 357 square kilometres of salt marsh shoreline in the upper Bay of Fundy (once called La Baie Française, its Acadian name), ninety percent was eventually dyked. The distinct identity of the Acadians came from their dyking techniques and their reliance on the Fundy marshlands. They likely brought their technology (introduced to the La Rochelle region of France by the Dutch) with them from France.

Dyke building began with the positioning of logs into the marsh mud in a wedge shape. Marsh mud and spruce brush were then packed around them and the structure faced with bricks cut from the salt marsh sods. The sods' knotted, packed roots held together well against the force of the incoming tides. Installed at the foot of the dyke, les aboiteaux prevented the sea from flooding the land at the same time as they allowed freshwater to drain into the sea. Basically, they were a form of hollow log sluice with a one-way flap made of hardwood at one end. The flap allowed rainfall and freshwater from the rivers to flow outwards but blocked the sea water from running up the rivers and into the dyked areas at high tide. No fertilizers were needed because nutrients were constantly replenished by the tides. By 1713, the Acadian population had reached five thousand people, and they harvested over twelve thousand acres of dyked marshlands.

In 1749 the English established a very large settlement in Halifax and thus began intense efforts to dispel the French from the New World. The Golden Age of Acadia, as it has been called, ended in 1755 with the Expulsion or Grande Dérangement. The British deported Acadians to the thirteen colonies along the east coast. Some fled to Quebec, others hid with their First Nations allies. Those who made their way south eventually found a new home in southern Louisiana amongst the French and Spanish Catholics. They settled along the waterways and swamps of the backcountry to the west and south of New Orleans. Their customs, and especially their cuisine, blended with Creole, which was itself a blending of Spanish, French, African, and indigenous American Indian elements. Acadians became Cajuns (a mispronunciation of their name) in their new homelands. When the Acadians were allowed to return to the Maritimes from exile, they found that English settlers had resettled their farms. Out of necessity, they learned fishing and commerce to replace the dyked farming that had once been their mainstay.

Acadians came from an agricultural background. In their new homelands, they cultivated English hay (mainly timothy and clover) as a main crop on the marshlands. It was used as mattress stuffing and to make mats. Spartina grasses were harvested in the fall as "salt hay fodder." Salt hay supported the horses that worked the Acadian marshland farms; it was sent to the mines and lumber camps on which the region's economy was built, and was shipped to New England and Newfoundland as an important export. Acadian crops transformed the salt marsh, including wheat, peas, oats, corn, barley, beets, parsnips, carrots, onions, chives, salad greens, cabbage, turnip, flax, and hemp. Flax, which supported a spinning industry, was used to make bedding and summer clothing. Hemp was used for grain sacks and some clothing.

As well as using plants they were familiar with from their homelands in France, Acadians used many locally available plants. Some they cultivated in their gardens, others they gathered wild. Cranberries and blueberries were widely used. In summer, two marsh plants, goose tongue and samphire, were harvested as greens. Acadians also gathered wild onion to make into soups and stews, added the leaves and blossoms of clover to salads, and made teas from the leaves of strawberry, raspberry, and mint. Clover honey was often added to remedies.

English, Irish, Scots • Between 1776 and 1783, British Loyalists fled the United States, most taking refuge in what is now New Brunswick, where they joined English settlers who had arrived in 1759, four years after the deportation of the Acadians. Another wave of settlers came during the historic Yorkshire Emigration of 1772–1775, when more than a thousand people left Yorkshire, England, to move to the Bay of Fundy. They found storm-flooded dykes, however, and their farming knowledge, developed in the uplands of Yorkshire, did not equip them for marshland farming. They learned how to repair the dykes from Acadians who had been left behind, confined to the English forts. Subsequent farming of the marshlands fell to Yorkshire and German settlers, but by World War One and its mechanized technology, the salt hay industry began its gradual decline. Tractors could not be used because of the drainage ditches that criss-crossed the marshes, so by the depression years of the 1930s, much of the dykeland was failing. Today, agriculture, primarily beef cattle farming, has taken over in the region, such that only about fifteen percent of the original salt marsh of Fundy is still connected to the sea. The Canadian Wildlife Service has saved some of the salt marsh as wetlands and wildlife sanctuaries.

In Britain, people historically made extensive use of wild plants for food and medicine. This knowledge was passed down through the generations and would have accompanied the settlers who moved here. Old cookbooks from England, Ireland, and Scotland give many recipes for the use of wild plants such as nettles, docks, watercress, raspberry, strawberry, and others that originated in the Old World and made their way to the New. Plant knowledge certainly passed back and forth between immigrants and First Nations, as it did among settlers. Irish and German settlers probably introduced potatoes and tomatoes to the Acadians, for example. Medical doctors and veterinarians (sometimes one and the same person) soon incorporated plants such as chamomile, strawberry leaf,

Dock

and plantain into their medicine kits in the early years of European settlement.

Germans • Germans brought with them an extensive background in gardening and farming, as evidenced by their farms around Aulac, New Brunswick, and Tatamagouche, Nova Scotia. German fairy tales tell us how important their traditional crops are to German culture. The famous Rumpelstiltskin story tells how hay was spun into gold. Another story, "The Spindle, the Shuttle and the Needle," alludes to flax similarly being turned into gold.

Potatoes, rye, and cabbage sustained German communities in the early days and today remain important crops in the region, as do apples, grapes, and blueberries. Wine, beer, and schnapps are made on several farms in New Brunswick today, and roasting chicory root as a coffee substitute is still an important German tradition. Called *bluemchenkaffee*, or small flower coffee, chicory root coffee is available commercially, or you can make your own by roasting roots gathered in the fall and running them through a coffee grinder.

One of the most famous German foods is sauerkraut, and like many other foods, it was often treated as medicine. Captain Cook, for example, attributed the scurvy-free health of his sailors to the high levels of vitamin C found in the sauerkraut they ate. Today, Germany has become well known in the field of alternative health care.

The Aboriginal Peoples of the Northeast

The Northeast is the aboriginal homeland of the Mi'kmaq, Maliseet-Passamaquoddy, Beothuk, Innu, and Inuit. The Mi'kmaq, Maliseet-Passammaquoddy, and Innu are similar culturally and linguistically. The Beothuk may have spoken a language in the same linguistic family but researchers are unsure. The Inuit language belongs to a different language family entirely.

Erroneously called "Indians" by Christopher Columbus in 1492, the aboriginal (meaning original) or indigenous peoples of the Northeast now refer to themselves collectively as First Nations. The Arctic Peoples refer to themselves as Inuit, meaning "the people"; Innu means "human being."

At the time of contact with Europeans about five hundred years ago, Aboriginal Peoples had sophisticated languages and cultures, rich cosmologies, and complex land and resource management systems. Their pharmacopeia incorporated a wide variety of plants used to treat the mind, body, spirit, and emotions.

Like the lands they live in, their cultures are varied and complex. Most have relied on gathering, fishing, and hunting, and at times on horticulture. Aboriginally (i.e. traditionally), they were shamanic peoples whose daily life was intricately interwoven with the seasons, the land, and its resources. Travel was based on the movements of the sun and the stars, and their seasonal round followed the available resources of game, seafoods, and plants.

Like other shamanic and aboriginal peoples worldwide, a First Nations relationship to the land is a sacral one. Traditional beliefs and practices about the management and conservation of plants are highly pragmatic but they emanate from a land-based spirituality and way of life. All creation is sacred. Plants, animals, birds, fish, insects, winds, mountains, echoes, and even a footprint or animal track are considered to be

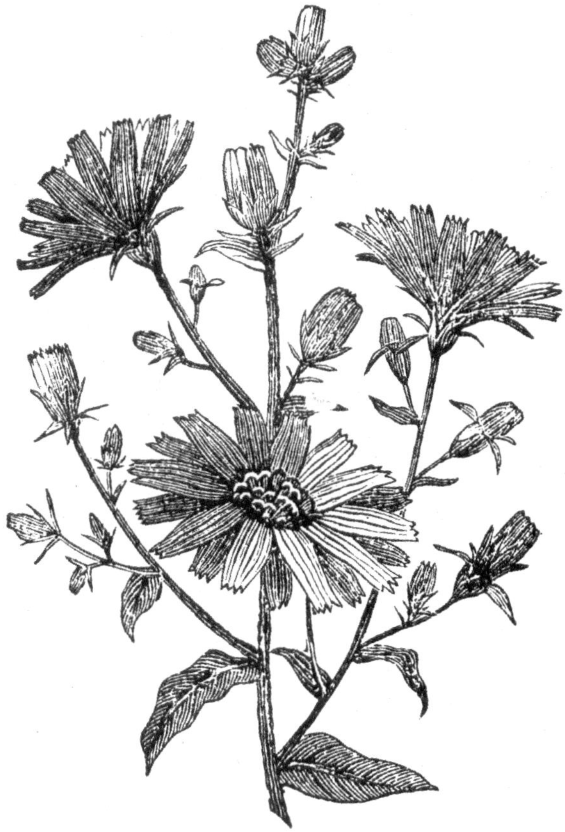

Chicory

	HOMELAND	LANGUAGE
Mi'kmaq	Maritime Provinces and Gaspé Peninsula	Algonquian
Maliseet- Passamaquoddy	New Brunswick, Quebec, and northern Maine	Algonquian
Beothuk	Newfoundland	Uncertain—Algonquian or separate linguistic family
Innu	Labrador and eastern Quebec	Algonquian
Inuit	Labrador, the Canadian Arctic, Arctic Quebec, and Greenland	Inuktitut

Juniper

"relations." Thus, work with plants keeps the spirit of the plant (and the spirit of Mother Earth) in mind. Behind the visible plant, rock, or river lays its hidden and unseen but true nature—its power or essence.

Plant knowledge is thus based in an understanding of the two worlds—the one that is seen, touched, and walked upon, and the unseen spirit world that guides everyday life.

First Nations learn about plants and their uses from the plants themselves, from the land that sustains them, and from the elders who counsel every aspect of their lives. Plants provide food, medicine, and materials for shelter, clothing, transportation, and technology. As well, plants provide spiritual guidance and teach us how to live a proper life.

Before a medicine man or medicine woman goes out to gather plants, a plant such as tobacco, artemisia, juniper, or sweetgrass is burned for cleansing and purification—of mind, body, and spirit. The smoke clears the mind of bad thoughts and negative energies so that good work results. It also is said to please the spirits of the land and of the ancestors, and to connect this world to the spirit world.

Today, First Nations in the region, as elsewhere in the world, are experiencing a new-found respect for their knowledge, including traditional plant knowledge, that many people have forgotten or been forced to abandon. Medicine men and women are being asked to share their knowledge with other First Nations as well as with the non-Native community.

Beothuk • (Other spellings include Bethuk, Boeothick, and Beathook in the historical literature; Beothuk was the name Beothuk applied to themselves.)

We know very little about the Beothuk of Newfoundland from contemporary accounts, other than that they had early and frequent contact with Europeans and Scandinavians, probably as early as AD 1000. They were the first peoples of the Northeast to meet Europeans in the fifteenth century and they were the most affected by the surge of Europeans fishing the Grand Banks in the sixteenth century. For much of their history, however, they were more isolated than the other aboriginal groups in the region in the same way that Newfoundland is separated from the rest of the region geographically. Four word lists in their language, collected when Newfoundland was explored and settled, are insufficient to determine whether it resembles other Algonquian languages or whether it represents a separate linguistic family.

The Beothuk were linked to the mainland by trading relationships and possibly marriage ties to the Montagnais of Labrador, as well as trading relationships with the Mi'kmaq. By the sixteenth and seventeenth centuries, hostilities over the fur trade developed among the aboriginal groups in the region, and also between the Beothuks and the French traders. The French, in fact, paid a bounty for Beothuk scalps, since unlike other Aboriginal Peoples in the region, the Beothuk were not considered to have any economic value (in fact, they had refused to join in the fur trade). They were more than dispensable: they were a hindrance to economic exploitation of the region.

With increasing European encroachment for fish and fur, and an ensuing war with the Mi'kmaq, the Beothuk were driven farther and farther inland and their communities increasingly harassed. The forest fires set to clear the Newfoundland coasts and harbours destroyed their local food source and they were forced to retreat to the interior, radically altering their seasonal round and their ability to sustain themselves. The last of the Beothuk were forced from even the fringes of their traditional lands. In the early nineteenth century they abandoned Newfoundland, some perhaps joining relatives and trading partners among the Montagnais in Labrador. By the early nineteenth century, however, the Beothuk were extinct.

Very little is known about Beothuk plant knowledge or about their seasonal round, which, at the time of contact, encompassed about three hundred miles from the interior to the coast and back again. Because of the unusual ecology of Newfoundland, the Beothuk were nomadic, relying on a wide variety of resources from different locations at different times of the year. Birch and spruce barks were basic materials for houses, canoes, packs, dishes, boxes, buckets, and cooking vessels. Roots, probably spruce, provided thread and twine. Pine arrows were shot from bows made of ash and fir. We know that they built birchbark canoes, and lived in conical pole-framed buildings covered with sheets of birchbark or caribou skins. Like their neighbours, they used traditional medicinal sweat baths for curing ills. These were built from hemispherical frames covered with skins; water was poured on hot stones to produce steam. They used red ochre to decorate their bodies, tools, clothing, and other objects. (They were the first to be called "Red Indians" because of it. The name was then applied erroneously to all aboriginals in the Americas.) Edible roots and inner barks were part of their diet, as were shellfish and birds' eggs, which were boiled in birchbark vessels.

Geometric designs of human figures, animals, and flowers were bitten into birchbark. Some of these were demonstrated by Shanawdithit, a Beothuk woman captured in 1823. She also made drawings of her people's way of life before she died in 1829, probably the last Beothuk in Newfoundland. The Newfoundland Museum in St. John's houses Shanawdithit's original drawings.

Inuit • The traditional Inuit homelands are the far north of the Canadian Arctic and Greenland with smaller numbers in northern Quebec and Labrador. The Labrador Inuit

Mountain cranberry

lived primarily on the coast, hunting seals, walruses, and water birds, and fishing on the rivers in summer. Barren ground caribou were the other staple of Inuit life. Aboriginally, they were known to visit the north coast of Newfoundland. Labrador Inuit settled farther south than any other Inuit group and thus were able to take advantage of the large trees that grew along the coast—wood was a scarce resource for other Inuit—for housing, implements, and utensils.

Labrador Eskimo, as they were called at the time, first made contact with Europeans in the 1600s with Basques from Spain who were searching for new whale-hunting opportunities. Once Labrador was transferred to Great Britain in 1763, New England whalers took over the previously French outposts. The British government supported Moravian missionaries by giving them land grants to aid in conversion of the Inuit to Christianity.

During these times, Labrador Inuit maintained their nomadic land use patterns, hunting game throughout the year. They travelled in kayaks and *umiaks* made from skins and wood frames. Winter houses were snow huts or earthen huts framed from whale ribs. In milder seasons, skin tents were used. By the 1700s, smallpox had taken its toll and in the 1800s, influenza reduced the aboriginal population drastically. During the nineteenth and twentieth centuries, an increasing dependence on the market economy altered the Inuit seasonal round. In 1975, the Labrador Inuit Association was formed to protect and promote Inuit culture, language, and traditional land claims.

August was known as "berry month" by the Labrador Eskimo. Berries are abundant in their homelands, perhaps more so than in any other part of the region. More than twenty varieties were identified by the Inuit, among them six bush varieties—blueberry (*Vaccinium pennsylvanium*), Indian pear (*Amelanchier canadensis*), dewberry (*Rubus articus*), dogberry (*Ribes cynosbati*), squashberry (*Viburnum pauciflorum*), and wild cherry (*Prunus pensylvanica*). A favourite dish was blueberries mixed with seal oil and salmon spawn.

Of the creeping berries, Inuit favoured cranberry, shrub blueberry—or as it's known in the region, duckberry (*Vaccinium cespitosum*)—and crowberry. Cloudberries or baked apples were eaten either green or when they ripened into their characteristic beautiful salmon colour. Wild strawberry produced tiny but flavourful berries as a midsummer treat. Cranberry was particularly important for its high vitamin C content, making it useful as a preservative and as a medicine. Greens such as dandelions were eaten with meat when they first appeared in spring. In times of starvation, even mosses such as caribou moss were used. Tubers such as the "Eskimo potato" (a variety of red lily) were dug up with pointed digging sticks and hung on strings to dry in the sun. Medicinal plants included the two types of Labrador teas that grow there. *Ledum palustre* was used as a tea for reducing fevers, and tansy was made into a tea for colds. Seaweeds were also used as both medicine and food.

Tansy

Innu • The Innu of Labrador were formerly known as the Montagnais-Naskapi. Since the 1990s, they have been known as the Innu, meaning "human being" in their language. Their traditional homelands are the eastern part of Quebec and the Labrador Peninsula. While the Inuit were coastal peoples, the Innu were traditionally interior peoples. The northern-most group of Innu is known as the People of the Tundra. Here, their territories border those of the Inuit.

Prior to the nineteenth century, the impact of Europeans was minimal. Innu traditionally lived a nomadic lifestyle, living in tents and hunting caribou. Caribou provided the Innu with meat as well as clothing, tent covering, tools, and babiche for stringing snowshoes. Once trading posts were established in Labrador in the nineteenth century, the Innu were brought into the market economy and their land base was eroded. The Innu Nation was incorporated in 1977 to protect the rights, lands, and way of life of the Innu.

Seals provided food for the Innu, Inuit, and settlers in Labrador. Walrus, ducks, geese, sea birds (meat and eggs), and muskox were also utilized. When trapping was introduced, marten, fox, mink, otter, lynx, and weasels provided a living for Innu and settlers alike. The Innu, like the Inuit, took advantage of the large variety of berries native to Labrador. Blackberries were gathered for food and sometimes for use as a dye. Dyes made from birch or alders boiled in a pot of water gave caribou hides a reddish-brown colour. Partridgeberries were traded for potatoes sent from Newfoundland.

The Innu homeland is a diverse landscape of mountains, woods, snow, and a short-lived but glorious summer when the plant life explodes. It ranges from boreal forest to bogs in the southeast, tundra in the north (encompassing the Torngat Mountains of the north), and low-lying eskers and river terraces in the south. The plant life is as varied as the landscape, from dwarf black spruce to the little blue and white violets that are among the first flowers to come up through the snow in spring. These give way to bluebells, wild morning glory, and sweet smelling flax that grow quickly in the short summer season.

Labrador Innu probably taught settlers how to "braid the birch" (birch was not common on the coast of Labrador) into sewing or knitting baskets; the technique may have been brought in from Quebec. Four long, thin strips of birch were braided into strands, which were sewn into functional but beautiful containers. Birch shavings were used in place of matches (which were scarce) to light pipes and lamps from the open fires they cooked on in their tents. Birchbark was also an especially useful covering for shelters. It was said that bears and porcupines would never tear it off as they would tarpaper or other human-made materials because it smells like the woods around it. It also lasts longer than other kinds of bark. Juniper was cut for toboggans, and as the lives of the Innu became more and more settled, moss was used to chink the log cabins that replaced their traditional tents.

Mi'kmaq • (also named Migmagi, Mickmakis, Mikmakiques, and Tarrantine in the historical literature)

When Europeans moved into North America, the Mi'kmaq occupied a large area of about 130,000 square kilometres. This included what is now Nova Scotia, Prince Edward Island, part of the Gaspé Peninsula in Quebec, and most of New Brunswick east of the St. John River Valley. Cape Breton was considered to be the "head district." As early as the eleventh century, Mi'kmaq probably encountered Viking settlers and explorers, and they may have been the first North American aboriginals to encounter European fishermen in the fifteenth century. As a result of their first meeting with explorer Jacques Cartier in 1534, Mi'kmaq became closely bound to the French in the fur trade. Mi'kmaq writings survive today, as well as detailed descriptions of their culture from French missionaries. A Roman alphabet–based writing system of their own invention was later modified for print by the Jesuit Father Pacifique.

Father Biard's Jesuit *Relations* (1611–1616) provides the beginning of Mi'kmaq ethnology recorded by outsiders. The Mi'kmaq are often described in the Eurocentric literature of the time as subordinate to the French. In fact, they remained independent of the French, although at times they acted against the British. They signed the Treaty of 1752 with the British, which established a basis for peace and ensured for the Mi'kmaq "free liberty of hunting and fishing as usual." In 1982, the validity of this treaty was upheld by the Supreme Court of Canada.

The pressures on the Mi'kmaq during the historic period were tremendous. In the 1700s, the government of Canada launched a campaign of genocide supported by the

Time Chart of European/Aboriginal Interaction in the Northeast

1000 AD	Beothuk are in frequent contact with Europeans and Scandinavians.
Prior to 1492	Basques may have been fishing the Grand Banks but probably had little contact with Aboriginal Peoples.
1492	Christopher Columbus lands in the New World.
1497–98	Cabot's voyages to "New-found-land"; contacts between Europeans and Aboriginal Peoples of the northern Northeast begin.
After 1492	European fur traders and fishermen begin to exploit the vast resources of the region.
1500	Europeans are fishing off Newfoundland/ Labrador/ Nova Scotia.
1500–01	Portuguese explorer Gaspar Côrte-Real sails along the Newfoundland coast, landing in Labrador in 1501; records of his voyages refer to fifty-seven kidnapped Indians, one of whom had European items probably left by John Cabot's crew in Newfoundland. Basque fishermen begin annual visits to the coast.
1524	Jacques Cartier encounters peoples prepared to trade in furs; Giovanni da Verrazzano's voyage contains the most extensive early account of Aboriginal People of the Atlantic coast.
1536	Whalers are operating off northeastern North America.
1598	A French settlement is founded on Sable Island.
Early 16th C.	European fishermen are travelling each summer to Newfoundland.
1604	French colonists establish a settlement in Maine on the Sainte Croix River; it is later moved to what is now Annapolis Royal in Nova Scotia.
1606	The settlement of Msr. De Poutrincourt is established at Port Royal, New France, establishing trade relations and a point of contact for Jesuit missionaries.
1609	Henry Hudson explores the New England coast in search of a route to China, first trading with, then stealing furs from, Indians of coastal Maine.
1749	Halifax is founded by the British
1755	Britain deports Acadians.
1763	New France is ceded to Britain by the Treaty of Paris.
1776–1783	Loyalists (British supporters during the American Revolution) flee the United States, most taking refuge in what is now New Brunswick. They are the first wave of English-speaking people into what is now Canada.
1783	Control of lands south of Canada and west of the Mississippi pass to the new United States.
1784	First United Empire Loyalist refugees from the U.S. arrive in Saint John, New Brunswick.
1812	Britain and U.S. go to war.
1834	Slavery is outlawed in Canada.
1867	Canada gains independence (dominion status within the British Empire).
1871–1921	Canadian government signs eleven major treaties with Native Peoples who are moved to reserves.
1876	Indian Act is passed.

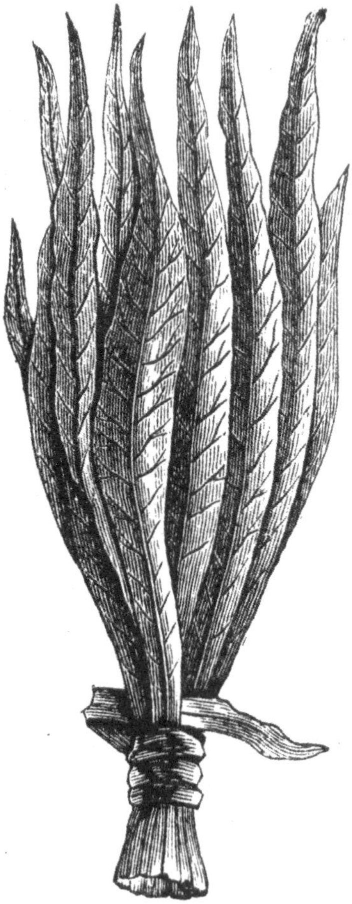

Tobacco

Christian churches. Starvation, small pox, hepatitis, pneumonia, and TB took their toll. The medicines that had served the Mi'kmaq so well for generations could not stave off the devastating effects of these Old World diseases. By the 1840s the population had fallen from between 10,000 and 35,000 to approximately 1,300. Much traditional Mi'kmaq knowledge, including plant knowledge, was lost along with the people that held it.

Like other First Nations, the Mi'kmaq had signed treaties of co-existence but did not agree to surrender their land. They received little or no compensation when their lands were turned over to settlers who were looking to rebuild their lives as farmers and business people in the New World. Mi'kmaq were moved onto about forty small reserves, most of which were inadequate for farming. By the early 1900s, most of the Mi'kmaq were settled on about sixty reserves, some with only a handful of inhabitants. By the 1960s, several thousand had migrated to the United States. Despite a declining population, Mi'kmaq continued to be politically important in the region. Today, about two thousand Mi'kmaq speakers live in Boston, a small number live in New York, and the majority, about six thousand, live in Canada.

Generally, the Mi'kmaq economy is described as maritime, with their seasonal round determined by changing resources. Aboriginally, Mi'kmaq spent part of each year on the coast, then moved inland in search of game. Spring and summer camps on the coast gave way to inland travel around September, the beginning of hunting season. In the spring, they fished for sturgeon, salmon, and other species. In summer they relied on shellfish, fish, and later, eels. Winter was a time when large game could be easily dragged back to camp over the snow. Through the winter, moose, caribou, beaver, and bear were hunted until the spring smelt spawn brought people back to the coast. After French colonization, Mi'kmaq began to settle for longer periods of time near the mission churches, although seasonal migrations continued for many well into the nineteenth century.

Aboriginally, Mi'kmaq cultivated tobacco and, after settling onto reserves (in Canada, reservations in the U.S.), became involved in growing potatoes, hiring out their labour for

Fiddlehead

the potato harvest. Mi'kmaq in New Brunswick travel south into Maine for the harvest to this day. Over time, plants introduced by Europeans became part of the First Nations pharmacopoeia; likewise, Europeans benefited from Mi'kmaq traditional plant knowledge. Today, fishing, hunting, and plant gathering continue to be important to the Mi'kmaq way of life and identity. Traditional medicines are still used by Mi'kmaq medicine men and women such as Jeorgina Larocque and Émile Gautreau.

Traditional Mi'kmaq territories were divided into seven districts, each politically independent. In one of these districts was located the petroglyph sites of what is now Kejimkujik National Park in Nova Scotia. It is an important heritage site for the Mi'kmaq, representing four thousand years of traditional use and conveying an unbroken sense of continuity with the past. The four sites there constitute the largest concentration of petroglyph images in eastern North America. The images, most of which date to the nineteenth century, tell the stories of the changing world of the Mi'kmaq amidst the increasing western presence in the region, including the steep population decline caused by the introduction of European diseases. Both Catholic and traditional spiritual images are carved into the slate rock. Curvilinear patterns that decorated the traditional peaked hats worn by Mi'kmaq women are there alongside the European-influenced Glengarry-style hats. Animals represented include caribou, moose, heron, porpoise, bear, deer, snakes, eels, fish, birds, and dogs. The most interesting petroglyphs for the purposes of this book are the references to plants. The double-curve motif, for example, represents the unfurled fiddlehead fern.

Archaeologists have documented Mi'kmaq use of strawberry, raspberry, bramble, chokecherry, sorrel, and mint prior to contact with Europeans. Berries were pounded and dried into rounded cakes. Roots and nuts were gathered. Birchbark was widely used for canoes, moose callers, dishes, and other containers. Birchbark boxes were beautifully decorated with quills, a technique that may have been learned from the nuns. Spruce root was also coiled into baskets, but the split-wood baskets for which the Mi'kmaq and Maliseet-Passamaquoddy are famous are of later origin. Certainly the blacks are known to have taught the Mi'kmaq how to make baskets in this style.

Daniel E. Moerman's *Native American Ethnobotany* provides the most comprehensive gathering of plants compiled from the anthropological and botanical literature known to have been used by the Mi'kmaq, plants they used for drugs, foods, fibre, dyes, fuel, cooking utensils, tobacco, and technology. My own research and that of Mi'kmaq communities themselves continues to expand the Mi'kmaq ethnobotany. New plants are always being added as plant knowledge is shared with healers, artisans, and gardeners from other traditions.

The Maliseet-Passamaquoddy • (In the historical literature, Maliseet are also referred to as Amalecites, Milicite, and Melicite; the Maliseet peoples' word for themselves is *Wolastoqiyik*, meaning the people of the good river and referring to the St. John River, Wolastoq. The Passamaquoddy are also referred to as Pesmocady.)

Traditionally, the Maliseet inhabited parts of New Brunswick, Quebec, and northern Maine. Their population was never large, so European diseases and warfare impacted them severely. Early contacts took place between Maliseet-Passamaquoddy and Basque, Breton, Portuguese, and Norman fishermen a century before the first written records were made. Seventeenth-century accounts include descriptions of contact with French explorers such as Champlain in 1603—they may have been the "Etechemins" named by him. John Gyles lived among them as a captive and wrote an account of Maliseet-Passamaquoddy life at the end of the seventeenth century. Allied with the French, many Maliseet were displaced by the English after the American Revolutionary War in order to accommodate Loyalists from the United States. By the twentieth century, Maliseet communities had developed in Connecticut, New York, and Massachusetts.

The Passamaquoddy are closely related to the Maliseet; in fact, they speak mutually intelligible dialects of eastern Algonquian. In many historical accounts, neither group was separated from the Mi'kmaq. Traditionally, their territories lay in the valleys of the St. John and St. Croix rivers, but Maliseet-Passamaquoddy identity and traditional land use has been confused by the United States-Canada boundary established in 1842. The two groups differed according to their livelihoods—Maliseet hunted inland whereas traditional Passamaquoddy territory was coastal New Brunswick and Maine, where they hunted sea mammals in the Bay of Fundy and gathered lobster, clams, and seabird eggs in the neighbouring islands.

Soon after the Treaty of Paris (1783), Americans began to pressure the Passamaquoddy to convert to Christianity. Assimilationist policies pressured them to adopt farming as a source of livelihood such that after 1842, the Passamaquoddy gradually moved to the United States and came to occupy reservations on ancestral lands in northeastern Maine. The Maliseet remained primarily in Canada. In 1980, the Passamaquoddy joined with the Houlton band of Mi'kmaq as well as the Penobscot in one of the earliest successful land claims suits, which was filed against the state of Maine.

Like the Mi'kmaq, the Maliseet-Passamaquoddy relied on hunting, fishing, and gathering, and followed a seasonal round based on available food resources. In summer, they lived together in settlements, travelling between their fishing sites and garden plots, gathering roots and wild grapes. In winter, they dispersed into smaller camps. Some Maliseet returned from winter hunting to plant corn, grains, and potatoes in the spring. In spring, fiddleheads were plentiful and were eaten as well as sold. In summer, the Maliseet-Passamaquoddy made baskets, in late summer they raked blueberries, and in the fall, they dried corn. They made canoes from birchbark—as did the Mi'kmaq—moosehide, and spruce bark, and today they are still known for their canoe building as well as their woodworking.

Métis • Derived from an old French word meaning "mixed," the name Métis is applied to people of mixed Native and European ancestry. Métis communities developed in what is now the Maritimes, Quebec, and Ontario by at least the mid-1700s. The organization of the western Métis and the ensuing rebellion led by Louis Riel is well documented, but parallel developments in the east are lesser known. Today there are Métis councils based in the Maritimes that maintain the rights of Métis to harvest traditional resources for sustenance, social, and ceremonial purposes.

Many Acadians claim mixed French-Métis ancestry, as ties between the Acadian and First Nations communities go back four hundred years or more. English-Native ancestry is less common because English settlers tended to arrive with entire families, while the French arrived mostly as single males who came as fur traders and explorers. Métis traditional plant knowledge therefore draws from both sides of their heritage.

One of the region's well-known Métis, Raymond Hébert, was a salesman from Barachois, New Brunswick, who made his own medicines for humans and horses, peddling his medicines, including rubbing compounds, throughout the Maritimes. Up until the 1940s, people in the Northeast relied

Mullein

on medicines made from the plants growing in the region.

Today, Métis such as Mary Lou Parker, Émile Gautreau, Blair Léger, Roland Surette, and others are celebrating their Métis heritage. They are taking care of the medicine plants they share, preserving and restoring the plants' habitats, practising their spirituality, and passing on their traditions to the next generation.

First Nations Traditional Plant Knowledge

Waves of settlers from Europe and Britain introduced many new plants to the region. The seeds of dandelion, mullein, and other introduced species travelled in bales of animal feed that accompanied the new settlers on their ships or in the dust in their pockets. Still other species were introduced as crops. Many of these new species became naturalized, quickly finding places in a landscape churned up by exploitation and settlement. Plants known for their medicinal properties soon found their way into the First Nations pharmacopoeia.

Some historians have claimed that pre-contact Mi'kmaq did not cultivate any crop except for tobacco. (This would have been *N. rustica* in earlier times, replaced by the commercially available *N. tabacum* in the nineteenth century.) Others describe a more complex relationship with plant resources before and during early contact. Recent archaeological evidence provided by Kevin Leonard's unpublished Ph.D. thesis, "Mi'kmaq Culture During The Late Woodland and Early Historic Periods," suggests that prior to contact, Mi'kmaq were already making a gradual transition to food production (although wild plants still fulfilled the bulk of Mi'kmaq plant requirements aboriginally). For example, they had begun to grow the Canada plum, *Prunus nigra*.

The primary source of information on aboriginal use of plants is, of course, the Aboriginal Peoples themselves, but since theirs is, until recently, an oral tradition, one looks for contemporary written records in the accounts of early European visitors. The writings of explorers, traders, missionaries, and adventurers in this region go back as far as Giovanni da Verrazzano's 1524 visit to the Atlantic coast. Europeans were as varied in their backgrounds as they were in their perspectives on the peoples they encountered. Some wrote detailed descriptions of their experiences over long periods of time, providing a clear picture of the lifeways of the people they met at the time of contact. Some were trained scientists; many were simply keen observers. Chrestien LeClerq is known as a great historiographer of the region during his time in the Northeast, but other sources require more scrutiny. Some accounts are quite fanciful or were plagiarized copies of others' works. As a result, these accounts tell us about the people and places being written about, but also about the writers themselves—their backgrounds, motives, morals, biases, and attitudes.

For the most part, early writers were in the New World for purposes other than the documentation of peoples they encountered. The first chroniclers were explorers, fishermen, and fur traders whose priority was economic, and they were soon followed by priests and missionaries whose ultimate goal was religious conversion. Generally speaking, medicinal plant knowledge was associated with native religion and for the most part was disapproved of by the church. As settlers moved in, making farms where there had once been forests, they too introduced new ideas about the relationship between humans and nature.

For example, the European sense of ownership—of claiming a plant or place—was foreign to aboriginal cosmology, which was rooted instead in notions of stewardship. The aboriginal name for a plant generally refers to the plant's typical location or to its uses or qualities; English scientific names often refer to the name of the first person (usually male) to identify the plant. Following Linnaean nomenclature, for example, the pitcher plant was named *Sarracenia purpurea* after Michel Sarrazin (1659–1734). He studied botany in early Canada and sent many seeds and specimens back to the Jardin Royale des Plantes in France.

With some exceptions, aboriginal plant use did not capture the attention of early writers in the way other activities did, even though knowledge of local plants and their uses sometimes meant the difference between life and death. For example, Jacques Cartier's suffering crew was saved by a traditional remedy for scurvy—a tea made from spruce needles. But later, botanists, physicians, and naturalists who settled in the region did gradually build up a knowledge base of the fauna and flora. This new knowledge helped Europeans survive in an unfamiliar environment and was also incorporated into the European pharmacopoeia.

As newcomers gained experience in the New World and their reliance on the native people decreased, records of the traditional ways of the Aboriginal Peoples became fewer. Older uses of medicinal plants were often not effective against new diseases introduced from the outside, and as government medical services were established, they were set aside. In any case, the old ways had already begun to disappear. Aboriginal Peoples were experiencing profound changes in all aspects of their culture, including the traditional means of preparing foods and medicines. They spent more time hunting and trapping for trade and less time hunting, fishing, and gathering for themselves. They became more dependent on the posts and missions for food as well as ammunition and other trade goods. Flour, oatmeal, and tea became regular parts of their diets as the supply of wild game declined. When supplies were scarce, however, First Nations resorted to the old ways: imported tea was stretched or replaced by Labrador tea or birch fungus; store-bought tobacco was stretched with kinnikinnick.

But eventually, increasing sedentarization and the introduction of commercial potato farming into northern Maine and New Brunswick changed the aboriginal way of life irreversibly, as they began to participate in the potato and blueberry harvest as migrant labourers. During the severe unemployment that affected the Maritimes in the 1930s, First Nations were hit especially hard. They were encouraged to plant vegetable gardens for their own food supply and to grow potatoes as a commercial crop.

In recent years, the interest in aboriginal medicines has increased. Aboriginal Peoples worldwide are being recognized for their knowledge, and science is corroborating the value of many of the medicinal plants that Aboriginal Peoples have used for generations. Aboriginal communities everywhere are giving priority to reclaiming their heritage and passing it on to their children. Some of the early accounts written by newcomers can assist in the reclamation.

Among those early writers on plant life and plant uses in the region was Giovanni da Verrazzano. By all accounts he was the first European to sail the coast of North America from Mexico to Newfoundland, meeting "Indians" in several places, including Maine. His voyage opened up the coast to successive European visits. Like other early visitors, he wrote about the new-found-land in terms of the European landscape with which he was familiar:

> *The land is, in situation, goodnesse, and fairenes, like the other...but not so sweete, because the countrye is more northerly and cold. Wee sawe in this Countrey many Vines growing naturally, which growing up take hold of the trees as they do in Lombardie...We found also roses, violettes, lilies, and many sorts of herbes and sweete and odoriferous flowers, different from ours.*

Other early explorers also described the plant life they saw. Jacques Cartier writes of cedars, as he called them (either juniper or a species of *arbor-vitae* since none of the true cedars are native to Canada), yews, white elms, pines, ashes, willows, peas, gooseberries, strawberries, and raspberries. Samuel de Champlain noted blueberry, walnut, bloodroot (*Sanguinaria canadensis* L.), and mayapple (*Podophyllum peltatum* L.). Mi'kmaq ethnology began with the Jesuit *Relations*, especially the work of the missionary Pierre Biard. The *Relations* are an important source of information about aboriginal culture, including land use. Biard mentions groundnuts growing underground near oak trees, describing them as strung together like a rosary. They are described as tasting like truffles, only better. Chrestien LeClerq (born in 1641), a Recollet critical of the Jesuits' work, included plant names in the system of hieroglyphics he developed for the Mi'kmaq language. Sir Richard Henry Bonnycastle (1791–1847) mentions clover, dandelion, pitcher plant, hawthorne, and others in a later account of Newfoundland. Nicolas Denys mentions ash, beech, and raspberry. Sieur de Dièreville mentions maple and fungus and had the bush-honeysuckle genus (*Diervilla*) named after him, carrying the plant with him back to Tournefort in France. Peter Fisher describes how he ate fiddleheads, grapes, and even the leaves of trees to allay hunger when moose and fish were scarce. He even writes how some "weeds" that got mixed in with his fiddleheads poisoned several of his companions. He also mentions making maple sugar in the spring.

A botanist, Jean-François Gaultier wrote a four-hundred-page manuscript of the plants he encountered on his travels, and had the genus *Gaultheria* (wintergreen) named after him. John Goldie mentioned hawthorne. Campbell Hardy notes cranberry, groundnut, mayflower, and hemlock. Sir George Head mentions birch and leeks "shooting up out of the earth." Louis Hébert grew apples and grain, and as an apothecary, he must have utilized the many medicinal plants of the region. Baron de Lahontan mentions oak, maple, cherry, and maidenhair fern, which was turned into great quantities of syrup by the inhabitants of New France and then exported to France. According to Alice M. Coats' *The Treasury of Flowers*, quoting from *A Compleat History of Druggs* by P. Pomet (1712), the Canadian maidenhair fern (which was of better quality than that found in other countries) was "cultivated with great Care in the King's garden at Paris" and prescribed for coughs and consumption.

Among the few women who published early chronicles of plant life was Emily (Shaw) Beavan, whose book, *Life in the Backwoods of New Brunswick*, describes the bright, scarlet berries of mayflower or trailing arbutus (*Epigaea repens L.*) as like clusters of coral on the snow. Catharine Parr Traill's botanicals, *Canadian Wildflowers* (1868) and *Studies in Plant Life in Canada* (1885), are classics. They are invaluable for their information about plants whose range extends into the maritime region.

The edible plants of the Northeast take on a special meaning in the perspective of history. They served indigenous peoples and newcomers alike and many are freely available to enrich us today. Perhaps we can share the wonder of the early visitors at the new and unknown landscape they found themselves in. Just imagine sighting the first dandelion of spring after a long winter in the trading post with no fresh vegetables to eat. Even so lowly and commonplace a plant as the dandelion can seem marvellous again.

The Northeast Region Today

Newfoundland's forests are still primarily softwoods—black and white spruce, balsam fir—as they were when the earliest explorers landed here. The white pine that once grew in extensive stands has since disappeared, however, and the settled coastal areas are now bare of trees. Extensive burning by early settlers to clear land for agriculture has left treeless barren lands that can support only scrub for grazing animals, such as caribou, and ground birds, such as willow ptarmigan. Plants of the barrens and boglands such as lichens and mosses, blueberry, Labrador tea, and orchids are found here. In the southeast, the centre of human habitation, the land is tundra-like with shallow, acidic soils. Plants that I first came across in the Arctic and alpine regions of other parts of Canada are well-adapted to the harsh conditions that glaciation created in Newfoundland. In the non-forested areas, for instance, glaciation left thin soils, sometimes only half an inch deep. Other parts of the province are boreal, characterized by a band of forest that stretches across Canada and is home to plants I wrote about in my first book, *Harvesting the Northern Wild*.

Labrador's ecology is diverse, containing tundra, boreal forest, and wetland. Consequently, Arctic species of more northern climes grow here. Its climate is cold, a result of the ice carried south by the Labrador current. Icebergs, present all year, also affect the temperature. Much of the coastline is rock covered with lichens and mosses. Small timber grows in some of the sheltered inlets, and there is an abundance of small game and berries. Potatoes, cabbages, and turnips are grown in the protected areas

Goldenrod

in small home gardens. According to a study of satellite images by Global Forest Watch Canada, Labrador and the northern part of Newfoundland retain the only large intact forests along the east coast.

Prince Edward Island is known for its stone-free, silty, bright red soils (from iron oxide) and its magnificent beaches. The beaches resulted from the deposition of sediments picked up and moved about by wave and current action. First "discovered" by Europeans in 1534, Jacques Cartier described it as "the fairest that may possibly be seen," and he described the many plants he came across as he sailed along its northwest coast. Once dense farmland, PEI has been transformed by resource exploitation from dense forest to scattered woodlots and open farmlands. Today, farming is PEI's main industry, with hay, potatoes, strawberries, grain, and dairy among its most important products.

New Brunswick is Canada's only officially bilingual province, a tribute to the recognition Acadians have achieved in the region. New Brunswick is home to the Miramichi River, the greatest salmon-spawning river in the world prior to the arrival of outsiders. The St. John River Valley produces livestock and potatoes, with mixed farming in the northwest. Global Forest Watch Canada found that virtually no large forest landscapes remain intact in New Brunswick (or in Nova Scotia or PEI), meaning that industry has left no areas large enough to preserve all of the province's naturally occurring species and ecological processes.

Nova Scotia's soils support primarily forests. Many hopeful early settlements failed because the land proved unsuitable for agriculture, though early settlers tried hard, cutting trees and ploughing the soils. One of the few exceptions is the central Annapolis Valley, known for its apples. The province's original stands of white pine were cut first by the French and later by the British Royal Navy, leaving scrub and barren lands. Today, less than one percent of Nova Scotia's forests are more than eighty years old. Cape Breton is rich in beauty and coal but has never been primary for settlement. Bras d'Or Lake is an inland salt "sea," a living reminder of Cape Breton's ancient origins.

Maine, the only American state to be included in the region covered by this book, is similar to the Atlantic Canadian provinces with respect to landscape, history

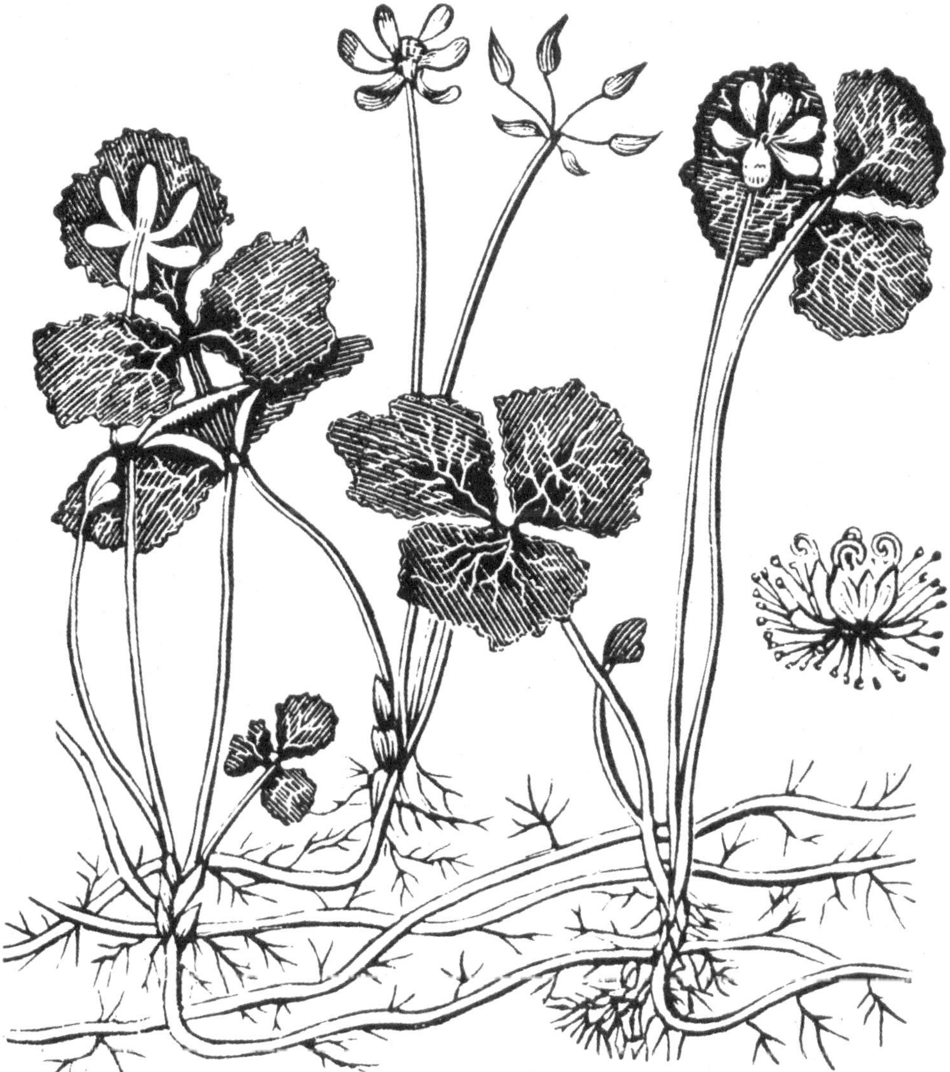

Gold thread

of exploration, and aboriginal cultures. Pine, spruce, and fir stands share the forests with sugar maple, aspen, and birch. Early European settlers were English and Scotch-Irish, and the Acadian presence in the road signage and gardens full of significant Acadian plants may be surprising to visitors unfamiliar with the history of the Atlantic region.

This is the route by which the expelled Acadians made their way into the American South, founding the Cajun culture of Louisiana. The Maine Department of Indian Affairs was established in 1965 to deal with the culture loss of the Maliseet-Passamaquoddy and to initiate community development programs based on Native American needs. Today, many Mi'kmaq and Maliseet travel into Maine to meet with friends and relatives for the traditional potato harvest and to participate in pow-wows.

A cultivated rose

The Northeastern Forests

The flora of the fall, comprising asters, golden rods and wild ever-lastings were all out, encircling the pearly grey rocks which strewed the barren, and every bush was wreathed with lines and webs of little spiders, marked by the myriads of minute dew-drops with which they were strung. Gradually warmed by the rays of the sun when, overcoming the surrounding barrier of the forest, they poured over the whole face of the scene, the little barren sparkled like fairy-land, the morning resolving itself into one of those glorious days for which the fall of the year is noted. Campbell Hardy: *Forest Life in Acadie,* 1869

Campbell's description of the beauty of the Acadian forest notes asters, goldenrod, and pearly everlasting, all native plants that still characterize the region today. Despite all the changes, the landscape of the Northeast remains a blending of boreal forest, mixed forest that combines conifers and deciduous trees, and deciduous hardwood forest. While old-growth hardwoods once covered much of the fertile land in the region, only a very few pockets are still standing.

The dominance of conifers is probably the main reason many writers have described the boreal landscape as monotonous. Yet there are enough differences in climate and geology to provide habitats for an astonishing variety of plants. The many species flourish in sites favourable to their needs and provide the careful observer with numerous rewards. Also called the Christmas tree or spruce-moose forest, the boreal forest is dominated by evergreen, needle-leaved trees (pines, hemlocks, spruces, firs, junipers) and deciduous, broad-leaved trees.

Generally, plant diversity increases as you travel south and decreases as you go north, where the stresses on plants become more severe. Fewer species have been able to adapt to the colder weather, deeper snow cover, shorter growing season, and relatively recent glacial retreat (only about twenty thousand years ago) in the northern areas. Plant colonization just has not maximized yet. The needle-leaved trees survive best in the north because their roots, stems, and needles are waxy from resinous chemicals that act as a sort of antifreeze. The shapes of the conifers, too, help them survive by minimizing snow buildup. Pointed, conical balsam fir and white spruce are more adapted to the cold and heavy snowfall than the spreading deciduous species.

Birch, poplar, and aspen grow where the land has been disturbed. Shrubby lowbush blueberry and herbaceous species include bunchberry, gold thread, and fireweed.

Lichens, the slow-growing but ubiquitous ground cover of the colder areas, are plentiful, while black spruce trees host hanging, straggly bearded lichen. Horsetails spread along the edges of ponds, lakes, and streams. Mosses, ferns, grasses, and sedges, such as white-tufted cotton grass, grow in abundance where habitats suit them. A few species, such as the pitcher plant, Labrador tea, cranberry, and some black spruce, can even tolerate the acidic waters of the bogs, where other plants cannot grow.

The northern hardwood forest includes yellow birch, sugar maple, pines, eastern hemlock, paper birch, pin cherry, mountain ash, and other species shared with the boreal forest to the north. Other trees and shrubs, such as flowering dogwood, red-osier dogwood, witch hazel, mountain laurel, and black cherry, are widespread throughout the boreal, hardwood, and transitional forests. The most obvious characteristic of the deciduous forest is the rapid change of leaf colours from green to yellows, tans, oranges, and reds as the growing season ends. As the daylight hours shorten, the temperature drops and the nights become cooler. Sugar maples turn from orange-red to yellow. (Maple leaves, suggests David Baker, M.D., show potential as an important and readily available antioxidant in the nutraceuticals industry.) Yellow birch turns from bright yellow to shades of yellow. The different colour patterns are a result of the varying chemicals and pigments that are revealed as chlorophyll (what makes the leaves green) breaks down.

Other plants also contribute to the Northeast's spectacular blaze of colour in summer and fall. The yellow pond lily forms a lush, floating carpet on the backwaters and shallows of lakes and ponds. Purply-red fireweed, a pioneer plant, transforms recently burned wasteland into swaying fields of colour. It blooms late into the year after many of the other wildflowers are gone. Some plants continue their display far past the summer months. Bearberries, cranberries, and rosehips are flashes of scarlet in the snow-covered winter landscape and provide essential food for wildlife throughout the long winter months. Even wolves and lynx rely on berries in times of starvation.

What Campbell Hardy wrote more than 150 years ago can enlighten our lives today:

A large bear at the end of the fall will weigh five or even six hundred pounds; this has been increased in domesticated specimens by oatmeal feeding to over seven hundred. Having awoke at last, the genial warmth of a spring day tempts him forth to try and find something to appease the growing cravings of appetite. What is the bill of fare? Meager enough generally, for the snow still covers the dead timber (where he might find colonies of ants), the roots, and young shoots and buds; but he bethinks himself of the cranberries in the open bogs from which, unshaded by the branches of the dark fir-forest, the snow has disappeared, disclosing the bright crimson berries still clinging to their tendrils on the moss-clumps and rendered tender and luscious by the winter's frost. Campbell Hardy: *Forest Life in Acadie.*

I hope you will enjoy becoming familiar with the plants detailed in the field guide that follows. I hope, too, that you will use these resources wisely, in partnership with all the other forms of life that rely on them.

Part Two: Field Guide

Bearberry (or Kinnikinnick)

Arctostaphylos uva-ursi (L.) Sprengl. • Heather family

Other names • mealberry, sandberry, mealy bearberry, upland cranberry, uva-ursi, stoneberry, hog cranberry

The name comes from the Greek *arktos* (a bear) and *staphyle* (a bunch of grapes).

Where found • Bearberry is wide-ranging, from Eurasia into North America; from the Arctic and Subarctic south into New England.

Habitat • It grows close to the ground on dry, sandy soils, gravelly ridges, and riverbanks, and often forms the ground cover in coniferous forests.

Description • The kinnikinnick shrub is low-trailing and spreading, reaching about 15 cm (6 in) in height. The stems are covered with white, soft hairs. The leaves are small, shiny, waxy, evergreen, and paddle-shaped. The bark is reddish with a rough, shredded texture. Tiny bell-shaped pinkish or whitish flowers grow in clusters, blooming in spring and followed by berries in summer. The berries are small, bright red, and contain tiny, hard "seeds." Berries ripen in late summer to fall, while the leaves can be used at any time of the year. Both can be retrieved from under the snow as they stay on the plants all winter and thus are an important food source for wildlife and an emergency food for humans.

As bearberry may be found growing among cranberry plants, the two are often mistaken for one another by the casual observer. Cranberry leaves and fruit are both smaller and shinier, however. One of bearberry's other names, hog cranberry, refers to the fact that it is used in similar ways to cranberry.

Usable parts • berries, leaves

Common uses • Raw bearberries may be mealy and tasteless but they are not "utterly inedible" as one early writer remarked. They are rich in vitamin C and carbohydrates. They can be used

fresh, frozen, or dried, though cooking the berries renders them more palatable. You can add them to soups or stews if you don't object to the small, hard seeds; or try boiling them first, then frying them in grease and sugar, a technique used by First Nations. Some people might prefer to sieve the berries after boiling them to remove the seeds.

Fresh or dried leaves can be used as a tea substitute or additive. Steep one or two spoonfuls of bearberry leaves in boiling water for about five minutes. The tea is pale green and mild-tasting.

Caution • Some sources warn that too many bearberries may cause constipation. In fact, bearberry has been used as a remedy for diarrhea. Early herbals warn that prolonged use of the tea can lead to stomach or liver problems. An excess can cause gastrointestinal problems, particularly in children.

Other uses • Dried bearberry leaves that have been rubbed into a powder have long been used as a tobacco substitute or additive. Bearberry is an important food source for birds and animals, including foxes, coyotes, and as the name implies, bears.

Medicinal properties • The leaves are rich in tannin, which has made them popular as a native tea. To make a bearberry tea, it is best to soak the leaves overnight in a quantity of cold water. This way, less tannin is dissolved and released than when tea is made with boiling water. As a medicinal tea, however, it should only be used for a few days at a time because of potential toxicity.

Bearberry also contains flavanoids, hydroquinones, and organic acids. Two other varieties of bearberry found in the United States (*coactylis* and *adenotricha*) were considered the most effective antibacterial herb for urinary tract infections until they were replaced by antibiotics and sulfa drugs. Today, bearberry is a recommended treatment for inflammatory conditions of the urinary tract. Bearberry is a bacteriostatic—it prevents the growth of some types of bacteria without actually killing them. In combination with dandelion root and horsetail extract, bearberry is used to assist urinary flow and to prevent fluid retention. It should not be used during pregnancy, lactation, or if an individual has a kidney disorder. Large doses may cause vomiting, according to one source.

I have seen whole bearberries in bear droppings in springtime. The bears were not digesting them, so perhaps they use them medicinally as a system cleanser or to get rid of parasites after a winter of hibernation.

In gardens • As an ornamental, bearberry's red berries and evergreen leaves look lovely all winter. The plant makes an exceptional ground cover and will thrive in poor soils or even sand. It is shade-tolerant and can even grow in high-traffic areas. Since bearberry roots form joints on the stem, it propagates where it is stepped on because this forces the stems into contact with the soil. I call it one of my kamikaze plants because it needs very little attention—no fertilizer, and very little, if any, watering. It is difficult to propagate, but try digging up frozen clumps in winter and transplanting them immediately into a bed made of four parts acid humus to six parts sand.

Blueberry

Vaccinium L. species • Heather family

The genus *Vaccinium* includes as many as four hundred species. Most of these grow in Asia, and include blueberry and cranberry. The *Vacciniums* are difficult to distinguish but all are similar in appearance and all are edible. I will mention the two most common types—the highbush and the lowbush. The wild lowbush blueberry (*V. angustifolium*) is the dominant blueberry of the region.

Other names • Common names can be localized and confusing; members of this genus (with the exception of mountain cranberry) are also called whortleberry, bilberry, bog blueberry, lowbush blueberry, dwarf bilberry, deerberry, whort, hurt, and huckleberry. (The name huckleberry is also given in southern regions to the similar, slightly taller shrub *Gaylussacia*.)

Cultivation of blueberries developed in the early 1900s with pioneers, who selected the largest wild plants to cultivate. Today, the North American wild, or lowbush, blueberry is grown commercially in Maine, Quebec, and Atlantic Canada, generating about $85 million a year in Canada and between $75–100 million a year in the United States. Europe and Japan are the major markets for Canadian berries.

Wild berries are healthier than the cultivated varieties, but some companies have been known to use pesticides on their wild berry crops. Most commercial crops are machine-picked when the berries have just turned blue even though they need a week to ten days to sweeten fully and taste their best.

Where found • Blueberries and their relatives are very common and widespread and are among our most easily recognized wild foods.

Habitat • Some grow in thickets, woods, cleared areas, or along side streams; others, in muskeg, peaty, or turfy areas; still others thrive on fire-blackened land.

Description • *Vaccinium* species are low or upright dwarf shrubs with deciduous leaves. Their bell-like flowers range from white to pinkish or greenish in colour. Flowers appear from May to July. Look for fruit in July on some plants and from early August through September on others. The leaves turn a beautiful red in the fall. The blue-black fruit of the wild blueberry looks about the same as the familiar cultivated blueberry, but is smaller. Wild blueberries are prized for their wonderful flavour, which is vastly superior to that of cultivated varieties. Both varieties are extremely versatile.

Caution • Black bears love both blueberries and fish. Since blueberries often grow along streams where bears are foraging, be careful in a blueberry patch. The noise from the water could make it difficult for you to hear a bear or for the bear to hear you. It's a good idea to make noise by singing, talking, or banging berry-collecting pails. And don't leave caught fish or their remains lying around your camp while you go off picking blueberries!

Usable parts • berries, leaves

Common uses • Blueberries are delicious fresh, dried, cooked, or frozen. They're great in pies, syrups, and jam, and can be added to bannock, muffins, pancakes, and your morning smoothie.

One way to clean blueberries is to drop them a handful at a time onto a tightly stretched woolen blanket laid out on a slant. Place a wide basin at the lower edge. The ripened berries will roll into the basin, leaving behind the leaves and other debris on the blanket while most of the unripe berries will bounce away. Just rinse the ripe berries, and you're ready to cook them or enjoy them fresh.

Both leaves and fruit can be used separately or in combination to make a very good tea. Just wash and steep the leaves and berries in boiling water. An old book of herbal remedies (an "herbal") recommends mixing equal parts blueberry leaves, thyme, and strawberry leaves—a very good mixture taste-wise. Herbals also recommend blueberry tea as a cure for diarrhea.

First Nations kept blueberries throughout the winter by digging a hole in the ground in a shady spot and waiting for the ground to freeze. Then they would put their berries in a birchbark basket or tin can, cover it, and place it in the "Indian ice box." Another traditional way to keep blueberries was by boiling them in animal fat or grease.

Other uses • First Nations traditionally made dye from blueberries for colouring hides and the quills of porcupines and birds.

Medicinal properties • Blueberries have become famous for their cancer-fighting compounds and antioxidant properties that neutralize the body's free radicals. In 1996, the United States Department of Agriculture (USDA) named blueberries among the most powerful antioxidants on the market. They have also been shown to inhibit cell damage and preserve memory. They are high in fibre and are known to help lower cholesterol. New research has found them helpful in the fight against heart disease and obesity.

Their time is ripe as a functional food and a nutraceutical. They are sold as juice and are being added to everything from dog food to food supplements. The USDA has even tested blueberry burgers—ground beef supplemented with blueberry purée—in elementary schools in Texas. Half a cup of blueberries a day is equivalent to the beneficial compounds to be found in three-quarters of a cup of strawberries or nearly three cups of corn.

In gardens • The commercial highbush varieties are easy to grow but they need acidic soil and thus grow well alongside holly, camellias, azaleas, rhododendrons, and other acid-loving plants. A soil pH between 4.2 and 5.2 is best (you can purchase inexpensive soil testing kits). Adding sphagnum moss, seaweed, pine needles, leaf mould, rotted oak leaves, sawdust, or wood chips will transform your soil from alkaline to acidic. Mix these into the soil, along with some sand, and test again until you have the right balance.

Blueberries make an excellent garden ornamental. In the fall their leaves transform a field into a glorious carpet of scarlets and russets. (I look forward to an annual autumn drive around Amherst, Nova Scotia, past an old, white farmhouse perched on a reddened hill of blueberry plants.) The highbush varieties can be trained into a tall hedge or living fence that can sustain itself for three or four decades given the right conditions. Just remember to plant at least two or three varieties for cross-pollination.

Since the hair-like roots are fine and fibrous, the plants should not be allowed to dry out, but they don't like to be waterlogged either. They need a loose, porous soil with lots of tiny passageways for their roots to grow and spread. Plant them in full sun in an open area. The best time to transplant from the nursery or another location is in the early spring or late fall when they are dormant. Blueberries need a good mulching over the winter to protect the plants and provide nutrients—use the same materials for mulch as you used to acidify the soil.

Bunchberry

Cornus canadensis L. • Dogwood family

Other names • dwarf dogwood, cracker berry, dwarf cornel

Where found • Common throughout the region; and from Labrador to Alaska and south into the United States; also eastern Asia.

Habitat • clearings, moist woodlands, mature bogs

Bunchberry is a woodland pioneer and indicator (abundant and conspicuous) plant of the boreal forest.

Description • Bunchberry is a low, mat-forming herb that spreads until it forms a lovely green carpet studded here and there with white "flowers." The carpet is made up of separate plants. Each has its own upright, wiry stem and grows to a height of not more than about 15 cm (6 in). The stem bears a main cluster of leaves near the top. Above the leaves there is a five-petalled white "star" that is easily mistaken for the flower. (Its petals are actually floral leaves.) At the centre of the white "stars" nestle the true flowers—they're very tiny and greenish-white or yellowish in colour. The tiny, berry-like fruits grow in clusters. They are scarlet and fleshy, and each one contains a single hard seed at its centre. The flowers begin blooming in late June. Berries appear from late summer through fall. A closely related species is similar in appearance and also edible: *Cornus suecica* L. has more numerous, smaller leaves, and purple flowers.

Bunchberry has the distinction of being the fastest mover of the plant world. Its stamens accelerate with a force greater than a rocket at a rate of 24,000 metres (99,000 ft) per second! This makes its movements faster than those of any animal. As the flower opens, it triggers the release of the four stamens, which shoot straight up and flick the pollen away from the flower. Perhaps this characteristic evolved to ensure pollen transfer by certain insects. Bumblebees, for example, can trigger the explosive mechanism, which then embeds pollen into the miniscule hairs on their backs.

Usable parts • berries

Common uses • Bunchberry is not commonly thought of as edible, but in fact the fruits are palatable and juicy enough, although some people find them mealy or tasteless. They're best eaten straight from the plant. When they are bright red, they are ready to eat. I enjoy crunching up the seeds between my teeth—they're a little like poppy seeds.

The hard seeds are a problem when you're cooking with bunchberries, but they can be strained out after boiling. Use the berries alone in puddings, preserves, or sauces for ice cream, custards, or meats, or mix them with other berries or fruits such as apples.

Calamus

Acorus L. • Sweet Flag family

Sweet flag was formerly grouped with another family, the Arum family. Now it is a separate family containing only two species (*A. americanus* and *A. calamus*) in a single genus.

Other names • sweet flag, muskrat root, flagroot, sweet myrtle, myrtle sedge, sea sedge, sweet cane, beewort, rédote, belle-angélique, sedge grass, cinnamon sedge, sweetroot, singer's root (in Mi'kmaq)

The name comes from the Greek "akoros," which designates a plant as having an aromatic root.

Where found • native to North America and Asia

From May through August, calamus may form large groupings where it is common. While its habitat ranges from Nova Scotia into Quebec and Ontario and south into parts of the United States, today it is found only in a few scattered locations in the Maritimes. Blair Léger, who took me to a patch in New Brunswick, said he had been looking for it for ten years before finding it there. It is also found in the Old World in temperate parts of Asia. In Europe it was cultivated for its aromatic root by at least the seventeenth century, thus settlers here may have either recognized it from home or learned its use from First Nations.

Habitat • sunny, moist places, especially marshes, wet meadows, and at the edges of calm water (such as ponds)

It may be found in coastal salt marshes just above the high-tide line and is often scattered amongst cattails or ferns or grows alongside them. The base of the plants are usually submerged into water. The patch I was taken to was growing in a dense carpet alongside a river. The ground was moist and since the area floods regularly, the calamus roots are continuously replenished by the deposited silt.

Description • A perennial, calamus grows up out of robust-looking rhizomes (the underground stems in which a plant stores its food reserves). Small rootlets extend outwards from the rhizome. Blair and I used a narrow rake to pull away the surface growth, and then dug into the wet soil at the base of the stems to uncover the brown and densely packed rhizomes. Following each one along its length with our fingers, we separated them out—some were several feet long, others just a couple of inches. Snapping them in pieces revealed a chalky white interior that tastes strong but pleasant.

Calamus is identifiable by its long (up to 2 m [6 ½ ft]), slender leaves that resemble iris leaves. Calamus leaves, however, are aromatic, glossy, and yellowish-green in colour. They are about a centimetre wide and are sharply pointed with sharp edges and a rigid mid-vein that extends along their length. Both the leaves and rhizome are noticeably aromatic with a tangerine-like scent but the rhizome has the stronger odour.

In spring, the new shoots—up to about 30 cm (12 in) high—can be harvested as a salad green. These were forming at the base of the stem when I saw them in late September.

Looking rather phallic, the fleshy, yellow-green cylindrical flower grows 5–10 cm (2–4 in) in length. It emerges erect out of the leaves about halfway up the stem, which grows to about 1.8 m (6 ft) tall. While botany books describe the flower as a way to identify the plant, the colony we found did not have a single flower showing—reportedly not uncommon. Later,

going through the rhizomes we had collected, we found a single, tiny flower, just over an inch long.

Usable parts • rootstock, young shoots; leaves (as floor covering)

Common uses • Many generations ago, calamus was a popular herb that was sold or traded widely throughout the region. Marc Lescarbot mentions seeing it at Port du Mouton in 1609. By the nineteenth century, calamus was commonly sold in the Boston market. In the early days of colonization here, it was exported to Europe for use as a medicine, insecticide, and for tanning leather, among other uses.

Calamus was well-known in North America for its uses in perfume. The root was chewed as a breath freshener and powdered into a snuff that was used to clear the sinuses. The root was candied as a sweet, and the tender inner parts of its young shoots were eaten raw in salads. As a condiment, it was used to flavour vinegars, beer, and other alcoholic drinks such as gin.

The rhizome has also been used as an insecticide. At the International Rice Institute in the Philippines, entomologists have studied the use of the oils derived from calamus in repelling insects, finding them useful in protecting stored grains.

Fresh or dried roots can be infused in oil, or a fresh root can be infused in alcohol, such as vodka, to render its scent. The rhizome needs to be chopped before drying or it will become too hard to cut. Cutting it into small pieces up to an inch long will allow it to dry evenly and will prevent mould from forming.

In Europe, the long leaves were strewn on the floors of homes and animal sheds as an aromatic floor covering.

Medicinal properties • One of First Nations' most valued medicines, calamus was named muskrat root due to the muskrats that nibble on its water-logged roots. In First Nations homes, the root was cut into pieces, strung up to dry, and hung throughout the house for later use. Calamus was chewed on by singers, allowing them to chant for hours at a time. It was used to treat sore throats, its numbing properties allowing one to swallow without discomfort. It was also given to children for toothaches, and chewed or smoked to break someone of the taste for tobacco. A small piece was even chewed to overcome mental fatigue.

The rhizome produces the drug calamus, which clinical study has shown to have antioxidant properties and antimicrobial activity. The plant contains an essential oil rich in asarone (related to mescaline), which has antibiotic properties but can be toxic in large doses. Calamus also contains minerals, starch, tannins, and a glucoside (that links to sugars).

Calamus acts as a diuretic. It was also recognized as carminative (expelling gas from the intestines) and was known for its use in stimulating and regulating menstrual flow and for promoting the appetite and aiding digestion. One of its most common uses was to treat serious coughs that couldn't be cured any other way.

Powdered, it was used as a snuff for nasal decongestion (in a dose of approximately 100–250 mg taken intranasally). The recommended dose for most ailments was no more than a teaspoon of the dried or powdered rhizome boiled for four to five minutes in 350 mL (1 ½ cups) of water.

Used as a sedative and for its calming properties in Ayurvedic medicine (the traditional medicine of India), the rhizomes were powdered and made into a tea. In Russia, festering wounds and ulcers were treated with a local bath of an infusion (in alcohol diluted with water) of the dried rhizome. In the Bible, named "sweet calamus," it is one of the components of ketoret, a ritual incense.

In small doses, calamus acts as a stimulant; in large doses, it causes vomiting. I was taught how to use it by placing a small piece of the rhizome—no more than half an inch to an inch—inside my cheek (under the tongue is too sensitive). I was told to bite into it occasionally and suck on it; then, after an hour or two, to take it out and keep it to use again later. I've seen it used this way in the Northwest Territories by the Dene, in central Canada by the Cree, and now have my own supply from a secret source in the Maritimes.

At one time, calamus was listed in the U.S. Pharmacopeia (1820–1916) and the National Formulary (1936–1950), but since 1968 it has been considered unsafe by the U.S. Food and Drug Administration. At least one credible source points out that the native North American variety of calamus does not contain any of the toxic compound found in other varieties. In Canada, under Health Canada's Therapeutic Products Compliance Guide, *Acorus calamus* may not be sold in any form.

Caution • In large doses, calamus can be toxic. I was taught to keep only a small piece of the rhizome in my mouth and for no more than an hour or two at a time.

Calamus resembles blue flag (*Iris versicolor*) when it is not in bloom. The rhizomes of blue flag are poisonous if taken internally (although it was once made into a candy). One way to distinguish them is by their leaves—calamus leaves are yellowish-green, glossy, and aromatic, whereas blue flag leaves are blueish-green in colour, dull, and odourless.

In gardens • Calamus has occasionally been grown as an ornamental in water gardens or in pots, but it likes rich soil. Calamus would be an excellent plant to establish in a pond or wet area, as it is disappearing quickly from the wild here. Be sure to buy from a reputable nursery that doesn't gather from the wild. Calamus is propagated from the seed, or more quickly, by its rhizomes.

Cattail

Typha L. species • Cattail family

Other names • quenouille

Where found • Cattails are surely one of our most familiar plants. The earliest date a cattail was recorded as native to North America was 1836. Just over twenty-two years later, it had become known throughout Canada and the United States. It spread rapidly as wetlands were disturbed or created by construction and development such as railways, roads, and waterways.

Habitat • streams, marshes, ponds, swamps, ditches

Description • Cattails are perennials. They grow in dense stands on erect stems that grow 120–150 cm (4–5 ft) tall. The leaves are long and linear and also erect. Tiny flowers are densely packed on thick spikes. In the fall, the tiny seeds roll off the dark brown spikes into the cottony wool that falls off in clumps. The wool can be seen floating on the water or carried on the wind.

Usable parts • young shoots, roots, leaves, root stalks, young flower spikes, pollen, seeds

Common uses • In the early days of New England, cattails were harvested for use as mats and baskets by First Nations. They were used to pad seams in boats and for packing material between barrels. First Nations in the region made flour, high in protein and carbohydrates, from the roots. The young shoots can be prepared like asparagus, but need a longer cooking time to make them tender. The starchy root stalks make a nutritious flour. The fluffy "wool" was used by First Nations as a natural "diaper" because of its softness and absorbency.

The lower parts of the leaves can be eaten as a salad ingredient, the young stems can be eaten raw or boiled, and the young flowers, or cattails, can be roasted. The yellow pollen can be added to a pancake or waffle recipe, imparting a beautiful golden colour. To gather the pollen in late summer, place a plastic bag over the cattail heads and shake. If the pollen is ready, you'll see lots of fine yellow powder spilling into the bag.

Cattails provide protection for wildlife and food for aquatic birds, amphibians, reptiles, and mammals. They are eaten by muskrat, geese, and beaver, and muskrats also construct their homes out of them. Red-winged blackbirds and other shorebirds make their nests among them.

Other uses • Soaked in oil or fat, they were once used as torches.

Medicinal properties • The Maliseet applied the leaves to sores. A First Nations remedy for an aching tooth was to rub it with juice squeezed from cattail stems.

In gardens • Cattails can be grown in swampy areas where few conventional garden plants could survive.

Chickweed

Stellaria L. species; *Cerastium* L. species • Pink family

Other Names • *Cerastium* species are also called mouse-ear chickweed. Various chickweeds are also known as adder's mouth, tongue-grass, starwort, and umbrella chickweed. The plant's name comes from the chickens and birds (especially quail, blackbirds, sparrows, and doves) that forage on its leaves and seeds. If you have chickens, chickweed from your garden makes a good natural feed. One of the most common varieties in our gardens and roadsides, *Stellaria media,* was introduced from Europe.

Where found • *Cerastiums* are found throughout the world as a perennial or annual herb. Chickweed is an abundant, hardy plant, easily recognizable and surprisingly useful. Plants that develop late in the season may stay green under the snow and get an early start in the spring.

Habitat • Common in fields, wastelands, woods, roadsides, and especially in the rich soil of gardens, where you may as well treat it as a salad green rather than a nuisance weed! If I have too much chickweed growing, I pull it out and drop it on my garden paths. It is soft to walk on and contributes to the mulch I build up to keep weeds down.

Description • Chickweeds are delicate-looking, with opposite, oval, untoothed leaves. (Mouse-ear chickweed, widespread in the region, has hairy leaves and stems.) The plant is small, but long-stalked and spreading, and grows upright or along the ground. Chickweed spreads quickly and can take over very large areas; it will often grow up a wall or a fence if there is no other room, or "down" a hanging planter or window box.

The tiny star-like flowers are white, usually with five petals, each of which has a single notch; they appear continuously through the growing season from spring to fall.

Usable parts • leaves, stems

Pick the young leaves and stems and remove the roots. The flowers can be left on. Note that later-season stems get too stringy to use. Always wash chickweed well, as sand and bits of debris get caught up in its spaghetti-like strands. I shake it out before I bring it into the house, then swish it around in a bowl of cold water to remove any undesirables.

Common uses • The leaves are excellent raw as a salad green. The young stems and leaves make a good potherb to add to stews—chop them up into bite-sized pieces. You can steam chickweed in a bamboo steamer or boil it for five minutes in very little water and serve as a cooked green. Chickweed makes a good, "wild" substitute for spinach in spanakopita or other recipes. The hairy leaves and stems of mouse-ear chickweed should always be cooked before eating.

Medicinal properties • Herbals have long recommended chickweed as a remedy for skin irritations. Bruise the leaves and mix them with lard to make an ointment for a traditional remedy. Or, instead of lard, use shea butter, extra virgin olive oil, or almond oil. Or you can simply apply the leaves directly.

Chickweed tea is recommended as a laxative. To prepare, crush the entire cleaned plant. Use one spoonful per cup, and steep in boiling water for five to twenty minutes.

In gardens • Chickweed will grow in your garden whether or not you invite it there, so you may as well benefit from its tenaciousness. (In Canada, *Stellaria media* is considered an agricultural weed.) You can extend its use as a wild garden green by snipping off the ends of the plant with a pair of shears or your fingertips; this will "force" it to keep growing and will inhibit seeding. Unless you want a bounty crop next year, however, pull up the plant before too much of it goes to seed.

Chicory

Cichorium intybus • Aster family (formerly the Composite family)

Other names • blue sailors, garden endive, succory, coffeeweed

Where found • Native to Europe and Asia, it is now widespread throughout the world and is common in the region.

Habitat • Chicory thrives in the wild along roadsides, in abandoned parking lots, and alongside outbuildings in weedy urban areas.

Description • Chicory is a branching, scraggly looking plant, growing up to 90–180 cm (3–6 ft) high, with flower heads that are a beautiful but unusual shade of blue (sort of a sky or periwinkle blue). It is unmistakable and especially stands out against the grey of gravel or stony soils. The flowers open only on sunny days. Its leaves are alternately spaced on the stem. At the base of the plant, the lower leaves resemble dandelion leaves. The leaves become smaller and sparser as they grow up the stem.

Usable parts • leaves, root

Common uses • The young leaves can be eaten in a salad or boiled as a vegetable.

Well known in Europe, chicory root is roasted, ground, and used as a coffee substitute. It is also mixed with coffee to impart a chocolaty, full-bodied taste.

Gather the roots in the fall, clean them well, and roast until dark brown in colour, then pound or pulverize the pieces in a coffee grinder. The root contains a compound called inulin that is converted, in roasting, to another compound that gives off a coffee-like aroma. Check out some well-known brands on the shelves of your local health food store. Look for brands that are organically grown. Some stores also sell it by mail order, including famous brands from New Orleans.

Chicory was cultivated in the Old World by medieval monks, then adopted by the Dutch and especially by the French. The French influence made its use commonplace in New Orleans, and then the Acadians expanded its popularity. I enjoyed a cup in a good French restaurant when I visited New Orleans years ago.

Other uses • Dr. Edward Bach recommended chicory for people who were overprotective and possessive. Bach made homeopathic flower essences in the 1930s to heal personality, mood, and emotional disorders, restore equilibrium, and replace negative emotions with positive. I have used the Bach Flower Remedies for many years and have found them helpful for humans and animals. They may be bought in small bottles in health food stores and are placed under the tongue with a dropper. I have also used Dr. Bach's system to make my own flower essences from locally available plants.

Medicinal properties • As a caffeine-free coffee replacement, chicory has lots of health benefits. It has been shown to lower both cholesterol and blood sugar (making it useful for those with diabetes). It may be helpful in weight loss and in stimulating the immune system, as well as alleviating the symptoms of intestinal diseases and harmful bacteria. It acts in similar ways to the beneficial bacteria found in yogurt.

In gardens • You'll find chicory growing wild throughout the region, but it also makes an excellent garden crop. Like other introduced weeds, it is very hardy and should do well just about anywhere in your garden. For plants such as chicory, why not leave a wild area in an undisturbed part of your garden? In the old gardens of Britain and France, this was where the fairies were left to carry on their lives undisturbed.

Wild Chives and Wild Onions

Allium schoenoprasum L.

Wild Leeks

Allium tricoccum Aiton • Lily family

Other names • Species in other parts of Canada are known by such names as nodding onion and wild garlic. Wild leek is also known as ramp.

Where found • Wild chives and wild onions are common and widespread throughout the region. Leeks, however, are rare delicacies today. In Quebec, the wild leek is a protected species under provincial legislation and may not be harvested for personal consumption in annual quantities of more than fifty bulbs or plants or more than two hundred grams of any of its parts.

Habitat • Wild onions and chives are found along riverbanks and lakeshores, in boggy meadows, and on rock slopes; wild leek likes rich moist soil and deciduous woods.

Description • From a distance, wild chives or onions resemble very dark green, spiky clumps of grass. Their clusters look just like the garden-variety chives. When you are close enough, their oniony smell is distinct and familiar.

The plants grow up to about 25 cm (10 in) high. The leaves are tender, cylindrical, and hollow. They are closed at the top and expand to encase the stem at the base. The roots are small bulbs that grow in a cluster, while the bright red six-petalled flowers of the native species look like pom-poms on their long, slender stems. (The introduced European chive we grow in our gardens has pink-purple flowers.) The flowers appear in June and July, and the leaves can be used throughout the seasons—you may even be able to find some still usable under the snow.

Wild leeks appear first as two or three shiny, light green, onion-scented leaves pushing their way up through the litter on the forest floor. The leaves are broad-shaped, and will grow to about 6 inches or more in height. They fall away before flowering, leaving a single flowering stalk that produces seed pods containing tiny, black seeds. The underground bulb is the important edible part, which can be baked or boiled like the garden variety leek, or pickled. A jar of pickled leeks is a wonderful gift to receive but it should be shared—better that everyone's breath smell the same!

wild chives

Caution • Be careful not to confuse wild onions with death-camass (*Zygadenus elegans),* also of the Lily family. While the two are somewhat similar in appearance, the wild onion is easily distinguished by its characteristic oniony scent, as is the wild leek.

Usable parts • bulbs, leaves, young flower heads

Common uses • Wild onion, chives, and leeks have an honourable and versatile role in our culinary history. Resourceful cooks will find all sorts of ways to use them fresh or cooked. I like to loop chive stems into a knot as a decorative edible on a salad or cheese tray. Chives and onion and leek leaves should be cut with sharp scissors rather than chopped to preserve maximum juice and flavour.

Chives are a lively garnish for such soups as tomato and vichyssoise. They can be added to salads, chopped egg sandwiches, mashed potatoes, scrambled eggs, or in any dish that requires a delicate onion flavour. Try adding chopped chives to bread dough when you are baking. The flavour of the bulbs is stronger than that of the leaves, and may be softened by boiling in several changes of water. The bulbs are delicious served hot with butter and may also be pickled.

Both the bulbs and leaves of wild leeks are edible and can be used similarly to wild onions and chives. They're very strong-tasting, however, so some people prefer them cooked. Eaten raw, the smell of raw leeks will stay with you for days.

Other uses • Herbals have long recommended chives as an appetite stimulant and aid to digestion. Since they contain less sulphur than domestic onions, they are more easily digested.

Medicinal properties • Since the times of early Egypt and the earliest Chinese dynasties, garlic has been considered as both a food and a medicine. In our own culture, garlic and onions have been the subject of many scientific papers. *The Commission E Monographs* (a German-established guide to herbal medicine) have supported their use for reducing risk factors associated with arteriosclerosis and specifically, coronary artery disease. Researchers at the National Cancer Institute in China have shown that members of the *Allium* genus can reduce the risk of stomach cancer. Dutch researchers have found that garlic has a beneficial effect on the reduction of cholesterol, but only in relatively high doses. Approximately four grams, or five to twenty average-sized cloves, per day for a 175-pound person are recommended for therapeutic use. Some people may be allergic to *Alliums* or report side effects such as heartburn. High doses should be used with caution by those taking anticoagulant drugs.

Fresh garlic has the ability to produce allicin, which, in turn, results in the formation of other active principles. Allicin has been shown to contain antiplatelet, antibiotic, and antihyperlipidemic activity and so is recommended as preventive medicine. Allicin is unstable in oil so minced garlic preserved in oil will probably not provide the same benefits. Onion shares many of garlic's properties, clearing arteries of fat that impedes blood flow.

In gardens • Chives are one of the easiest herbs to grow, inside and outdoors. A clump of garden chives can easily be divided and transplanted to another part of the garden or into a container. I plant them in clumps among my flowers. I keep a pot of them outside my kitchen door, bringing it in before the weather turns cold so I can snip fresh chives well into the winter months. A small pot does well inside on a sunny windowsill.

Cloudberry

Rubus chamaemorus L. • Rose family

Other names • bakeapple, baked apple berry, foxberry, maltberry, muskeg berry, Mars apple, Jonesberry, ground mulberry

In Scotland it is known as avrons, so it would have been a familiar friend to Scottish settlers in the region. In Norway, it is called molteberry; in Russia, maroshka.

As with many common names, there is some confusion about the term "salmonberry." Some writers apply the name to the cloudberry, but it is applied to different species as well, some of which do not grow in the region. For example, I'm used to calling cloudberry by the name salmonberry—that was the name I came to know it by in the Arctic. The ripe berry is a beautiful, distinctive salmon colour.

Where found • Widely distributed through the tundra regions from Greenland to Alaska, common on the coasts south into Maine but less easily identified inland where it may not set its fruit.

Habitat • moist areas of peat, turf, or muskeg

The beautiful fruit, which ranges from salmon to orange-gold in colour, really stands out against the dark soils of its preferred habitat.

Description • Cloudberry is actually a species of low or creeping raspberry; however, in appearance it bears little obvious resemblance to the more commonly known raspberry, which grows as a bush. Cloudberry, in comparison, grows low to the ground, from 5–15 cm (2–6 in) high. At first glance, the leaves look a little like strawberry leaves. Cloudberry leaves are rounded, mostly five-part, toothed along the edges, and somewhat leathery. They are firmer and tougher than strawberry leaves and have no prickles. Each plant bears a solitary, white, five-petalled flower and a single fruit. The fruit is a beautiful red colour at first, turning to salmon, then amber, then yellow when it is ripest and juiciest. It could be said to taste a little like apricot, but really, cloudberry has a unique and incomparable taste. Cloudberry flowers in June and July, and bears its fruit in August.

Usable parts • berries

Common uses • Cloudberries are one of the greatest treats to be found in the wild; however, this is one plant whose locations I keep to myself! The ripe berries are very soft and juicy, making them difficult to package or transport. It's tempting and sensible just to eat them where you find them. If you do happen upon more ripe cloudberries than you can eat at once and are able to take some home, enjoy them with cream or custard or make them into an uncooked fresh fruit pie. The fresh berries can also be frozen. You may want to make them into a freezer jam or they can be cooked and made into preserves, pies, or puddings. Newfoundland recipe books often contain recipes for cloudberry (which they call bakeapple) jam. Scandinavians market cloudberries commercially. In specialty shops, you may find them sold as preserves or bottled whole in sugar syrup or water.

Both Norway and Russia preserve the berries in alcohol and sell the liquor commercially. Here are directions for a homemade version of flavoured vodka: Clean the ripened berries but don't wash them. (You don't want moisture in the berries, which can produce mould and also dilute the alcohol.) Drop them into a bottle of good-quality vodka (drinking a little as you go to make room!). Cloudberries will keep for months this way, even longer. They infuse the vodka with their flavour and when the vodka is finished, the "boozy" berries make an excellent topping for ice cream or cake.

Other uses • The young shoots produce a grey or black dye.

In gardens • Unlike other wild berries, cloudberry has never become a common garden plant. In my travels, I have never seen it cultivated. It requires a lot of space for its long underground rhizomes to grow and prefers cooler, boggier conditions than most gardens can provide. If you do have a suitable area in your garden or woods, you can propagate it from seed or by transplanting parts of its rhizomes. Cloudberries are best picked in the wild, making them a precious find.

Clovers

Trifolium L. species
Melilotus Mill. species • Pea family

Where found • Clovers are widespread throughout the world. None of the clovers growing here are native. They are Eurasian species, introduced from Asia and Europe, and are now common and widespread weeds in the region, as they are in the rest of the world. The *Trifolium* genus includes about 250 species, and the *Melilotus* genus more than 20.

Habitat • common in fields, meadows, wastelands, gardens, and along roadsides

Caution • Although I have never experienced any difficulties myself, some sources state that clovers may present digestive difficulties and are thus potentially toxic to humans. Use the clovers with caution and in small quantities at first to see if you react.

Melilotus officinalus, if not dried properly, has been said to produce a potentially fatal mould. Take special care when drying this yellow clover and discard if mould appears.

Trifolium Species

Other names and varieties • red clover, white clover, creeping white clover, Alsike clover, rabbitfoot clover

Description • *Trifoliums* are the true clovers and the most familiar. They are unmistakable because of their three-part leaves or trefoils. The flower heads are red, pink, or white, appearing in July and August. Red clover has purple-red flower heads and is taller than white clover. Alsike clover is whitish or pale pink.

Usable parts • roots, young leaves, flowers, seeds

Common uses • Clover is not as pleasant tasting as many wild plants, but its abundance makes it important as a protein-rich emergency food. The roots can be scraped and boiled like carrots. They are a good source of carbohydrates and roughage, and therefore useful as a laxative.

Fall or spring is the best time to collect roots. The tapering anchor root is not the one you want; look instead for the single strands growing vertically in the top layers of soil. These look like twisted spaghetti, and indeed, a related species is called "Indian spaghetti" by the Bella Coola Indians of British Columbia.

The greens can be eaten raw in salads, but more than a few are difficult to digest. To eat them in quantity, boil for five minutes or longer. The very sweet white ends at the bases of the flowers are good to nibble. The flower heads can be used fresh; just steep them in boiling water. They can also be dried and made into tea, or added to other teas. To dry them, gather them on a dry day after the morning dew has evaporated, lay them out on a screen or clean cloth, and dry them slowly at room temperature, out of the sun.

Other uses • The flower heads and seeds can be ground into flour. The flowers may be made into wine. In medieval Britain, clover was dried and stored as a winter fodder for farm animals.

Medicinal properties • Clovers have long been important in European folk medicine. The flowers were made into tea to cure constipation and indigestion and to stimulate the appetite, or they were applied externally as a poultice to cure rheumatic pains, relieve inflammations, and heal wounds and sores.

Melilotus Species

Other names • sweet clover, melilot (from the Greek, *meli* meaning honey; *lotus* the name of an ancient plant)

Description • These white- or yellow-flowered clovers are upright, tall, and branching plants that can cover large patches of ground. They will be quite noticeable because of their height—up to 1 metre (3 1/4 ft) tall. The tiny flowers grow on elongated spikes, appearing from May to October. The leaves also grow in threes but are narrower than the *Trifoliums*, and when crushed, have a slight scent of vanilla. The seeds are pea-like.

Usable parts • young leaves, seeds

Common uses • Before the flowers develop, the young leaves can be used as a salad green or boiled for five minutes to be eaten as a cooked vegetable.

The leaves can also be dried, then crushed and added to baking or puddings for a vanilla-like flavour. The seeds can be added to stews and soups.

In gardens • The *Trifoliums* make excellent ground cover but not where you're going to be walking in bare feet: In the heat of summer, a patch of clover will be alive with bees! The bees will be making a lovely, clover-scented honey that is delicious in tea.

Cow Parsnip

Heracleum lanatum Michx. • Parsley family

Other names • Indian celery, Indian rhubarb, masterwort, cow cabbage

Where found • A common and abundant plant all over Canada, with the exception of the Arctic, and into the United States.

Habitat • open fields and along roadsides

Description • Cow parsnip grows up to 2 m (6 1/2 ft) tall. Its stem is stout, grooved, and hollow. The large lower leaves are divided into three irregularly serrated sections that resemble maple leaves. The upper leaves have obvious "wings" or papery sheaths on their stems. Older stems and leaves have a strong, rank odour. The taproot is thick and fleshy.

The tiny white flowers, while individual, are grouped together in flat-topped umbrella-shaped clusters that appear in early summer and can measure up to 20 cm (8 in) across. Several clusters may grow on one plant, with the one growing at the top of the stem being the largest. The seeds that follow may be gathered in summer, while stalks and stems should be harvested in spring when they are still young and tender. Roots may be dug in fall and also in spring.

Usable parts • young leaf stalks and young stems, roots, seeds

Common uses • Early explorers recorded how they boiled the tops and mixed them with dried meat for their suppers. They also witnessed First Nations roasting the tender tops of this plant over the fire and then removing the outer rind.

Pick the young leaf stalks and stems before the flowers expand. Peel them, as their "skin" smells strongly and may cause skin irritation. The peeled stems and stalks can be eaten raw, but boiling them in three changes of water is recommended, particularly if they are older. The roots, if young and tender, can be cooked like parsnips. The seeds can be added to soups or stews as a seasoning. Dry them and put them on your table along with pepper and salt to sprinkle on your food.

Medicinal properties • Herbals recommended a drink made from cow parsnip roots or seeds for colds, cramps, asthma, and colic. They suggest that a distillate can be used externally for sores and wounds.

Caution • The unpeeled leaves or roots of cow-parsnips can cause skin irritation in people who are susceptible. Wear gloves when gathering and preparing them.

Cow parsnip belongs to the same family as water hemlock (*Cicuta* species), and resembles it in appearance and habitat. Water hemlock is very toxic and can cause death or convulsions, diarrhea, stomach pains, and delirium. The rootstock in particular is poisonous, but so are the other parts of the plant.

Water hemlock grows in standing water or in moist habitats along stream banks, in roadside ditches, in marshes, and along lakeshores or creeks. Water hemlock plants are more slender than cow parsnip and have finely divided leaves, as opposed to the maple-like leaves of cow parsnip. Even botanists, however, cannot always differentiate between the two with certainty. If you are not absolutely sure of your identification, don't take the chance.

Cow parsnip also resembles another deadly poisonous plant, poison hemlock, *Conium maculatum* L. It grows in the region in dumps and waste places but is also found as a garden plant. Like water hemlock, the whole plant is poisonous, but the seeds are especially dangerous.

Cranberry

Vaccinium vitis-idaea L. / *Vaccinium oxycoccus* L. *Vaccinium macrocarpon* Ait. • Heather family

Other names • *Vaccinium vitis-idaea* is known as mountain cranberry, foxberry, wild cranberry, and alpine cranberry as well as lingonberry, cowberry, and partridgeberry. *Vaccinium oxycoccus* is known as bog cranberry and small cranberry. *Vaccinium macrocarpon* is known as American cranberry or large cranberry. This variety was enjoyed by the Pilgrims in New England, who probably called it "fenberry" after the wild cranberry that grew in England in swampy places or fens. In later years, fenberry was eventually renamed "kranbeere," from the Dutch. In 1677, Massachusetts colonists sent ten bushels, gathered from the wilds, to Charles II in England to appease his anger over their coining of "pine-tree shillings" without his permission.

There are more than one hundred varieties of cultivated cranberries; many of them are named for the families that grew them. Cultivation probably began in the early nineteenth century. They were among the first fruits to be canned.

Highbush cranberry (*Viburnum* species), a member of the Honeysuckle family, is unrelated.

Where found • Bog and mountain cranberry can most easily be distinguished from one another by the different habitats in which they grow. American cranberry has larger leaves and berries although it grows in similar habitats to bog cranberry.

Habitat • Bog cranberry prefers wet bogs and muskegs, while mountain cranberry grows in more open areas, on slopes, and in drier soil. American cranberry grows in the wild from Newfoundland to as far south as Tennessee but is particular about its habitat, choosing only certain lowland bogs.

Despite the name, the soil in a cranberry "bog" is actually dry and peaty and usually surrounded by water (you often find sand dunes nearby). A long trek through a bog can be tiring, though, because of the spongy quality of the soil. But then, its sponginess makes it a comfortable place to lie down—after you've picked the berries underneath you—and enjoy the piney, peaty smell of a cranberry bog.

Description • Both bog and mountain cranberries are low-trailing, evergreen shrubs. Their leaves are dark green, very shiny, and oval in shape, remaining green all winter. The stems are twiggy and slender, taking root every now and then. The fine, four-part bell-like flowers are a lovely pink colour, blooming from June through August. The ripe berries are red to dark red in mountain cranberry, reddish in bog cranberry, and red in American cranberry. They are very shiny, firm, and the size of garden peas; American cranberry has larger berries that are pink to beige inside. The berries appear from August to November and remain on the bushes throughout the winter.

Cranberries have air pockets that make them bounce like tiny rubber balls; in fact, they were once called bounceberries for this very quality. This is an advantage for the commercial harvest because the berries float. The best berries bounce the highest—up to 5–8 cm (2–3 in). As a result, you can easily separate your wild-harvested berries from their chaff by bouncing them around in a basket. Or try blowing away the chaff with your breath. If that gets too tiring, try using a hair dryer.

You will often find cranberry and bearberry plants growing together. Of the two, cranberry leaves and berries are the smaller and shinier, and cranberries are not mealy like bearberries. The cherry-sized cultivated cranberry is much larger and less tart than the wild varieties.

Usable parts • berries

Common uses • References to the wild cranberry are abundant in historical writings, and today they are still one of the most sought-after of the wild edible plants of the region. Going out for a hike on a sunny fall day to pick cranberries for Thanksgiving dinner is an annual tradition for many of us and, of course, the sauce tastes far better than any store-bought kind!

In the northeastern United States, Native Americans taught the Pilgrims about cranberries, and they became the first American fruit to be prepared in European kitchens. Cranberries also became invaluable to New England sailors for scurvy prevention, in the same way that the lime (the source of the name "limey") was used for English sailors, and sauerkraut (hence the name "kraut") for the Germans. Today over fifty percent of the commercial cranberry harvest in the United States comes from Cape Cod where, historically, farmers converted their wetlands into cranberry bogs. Today, the northeast United States is known for its cranberry harvest. Beginning around the third week in September and peaking by mid-October, it draws thousands of visitors on Columbus Day each year.

Wild cranberries stay on the bush all winter long, and in fact, their taste improves by the first frost. They are an important emergency food source when little else is available. They can be gathered frozen from under the snow and kept frozen until needed. The commercial harvest peaks in mid-October and is over before the first freeze (freezing air kills the plants; ice does not).

Commercially, they are harvested wet or dry. In a wet harvest, the fields are flooded. A machine called an "egg-beater" agitates the waters so that the berries fall off the bushes and float to the surface. From here, they are gathered with long-handled rakes into corrals made of boards, then pumped into trucks (although sometimes they are pushed onto conveyor belts and then into trucks). They are then turned into sauce or juice.

In a dry harvest, the best berries are sold fresh and loose or packaged for home use. A harvesting machine is driven across the bog to rake in the berries and put them into burlap sacks, which are then emptied into crates for airlift to trucks by helicopter. A few areas, especially along the edges of the bog, are still harvested by hand. Wooden cranberry scoops that have a wide-toothed rake to comb the berries into the scoop are the best way to harvest the berries. Wild cranberries can be easily cleaned by the same method suggested for blueberries.

Cranberries contain their own pectin and are rich in vitamin C. Wild cranberries can be used in any recipe calling for cultivated cranberries. They can be cooked as a sauce, jelly, or pie filling. They can be used fresh or frozen as an addition to bannock, stuffing, cakes, muffins, or pancakes. Or they can be pressed into an excellent cold drink and diluted with water, soda, or ginger ale. They make a delicious vinaigrette, crushed or juiced and added to cider vinegar and extra-virgin olive oil. They are one of the easiest berries to freeze for future use. I just drop them in a freezer bag, suck out the air through a corner, and twist-tie the bag to seal it.

Medicinal properties • Cranberries are saturated with vitamin C, making them an important nutraceutical. For years now, doctors have recommended that women drink cranberry juice daily to protect against urinary tract infections, and several times daily when an infection is present or threatened.

In gardens • The true cranberry (in comparison to substitutes such as the highbush cranberry, *Viburnum* species) is not impossible to grow in home gardens, but it is difficult to simulate the right conditions. Personally, I prefer foraging.

Crowberry

Empetrum nigrum L. (black)
Empetrum rubrum L. (red) • Crowberry family

Other names • curlewberry, heathberry, mossberry, crakeberry

The crowberry's scientific name comes from the Greek *em* meaning "upon" and *petros* meaning "rock" to describe its preferred habitat.

Where found • Crowberry is a wide-ranging plant throughout much of Canada, and is found as far north as the Arctic islands. In the Maritimes, both the red and black crowberry are native species.

Habitat • common in peaty, open soil and bogs, rocky or sandy places, and in open forest

Crowberry often forms the ground cover on north-facing slopes.

Description • Crowberry is a freely branching evergreen shrub that forms a low mat. Its needle-like leaves resemble juniper or spruce leaves. The flowers, tiny and inconspicuous, are green or purple in colour. They appear in spring as soon as the snow melts, lasting through July, August, and September. Each berry is about the size of a pea, and is juicy and shiny. Black crowberry is purplish-black with a whitish bloom; red crowberry has dark red berries. Unlike the black crowberry, the young twigs of red crowberry are covered in dense white hairs. The berries contain a number of large, hard seeds. Berries are borne from July on, and can be gathered through the winter.

Useful parts • berries

Common uses • Crowberries are well known and appreciated in northern Europe, but receive little attention in Canada. Don't overlook them, though, for they are hardy and versatile. They're abundant, wide-ranging in many places, and easy to pick. Pulling them off in handfuls will not bruise the fruit. Some people say the taste is too much like turpentine for their taste, but I would describe it as acidic and flavourful, although less extravagant than the blueberries or cranberries that often grow nearby.

Crowberries are good added to muffins or pancakes. You can use them fresh, frozen, or cooked with sugar and a little lemon juice to be served with ice cream or custard. They can be mixed with blueberries to give them more flavour. They also make good jelly or jam. One early source notes that the fermented juice produced a sparkling white wine.

Other uses • The fruit can be picked all winter (its flavour actually improves after freezing) and is thus an important emergency food. A handful of berries is an effective thirst-quencher if you've forgotten your water bottle on a long hike.

Medicinal properties • Crowberry was once a traditional treatment to prevent and treat scurvy.

In gardens • Crowberry flourishes in rock gardens or in shady, moist parts of a garden, preferring a pH level between four and five. It does well in the Arctic and the colder regions farther south as it does not like intense heat. To propagate, take cuttings in midsummer and place them in a cold frame in a mixture of peat and sand.

Currants

Ribes L. species • Gooseberry family

Where found • Species of the genus *Ribes* grow throughout the region, both native and introduced, and include currants and gooseberries. Among them are:

Gooseberry—*Ribes hirtellum* Michx.
Wild red currant—*Ribes triste* Pallas
Swamp currant or bristly black currant—*Ribes lacustre* (Pers.) Poiret
Skunk currant—*Ribes glandulosum* Grauer
Red currant—*Ribes sativum* Syme
Blackcurrant—*Ribes nigrum* L.
Eastern Black Currant—*Ribes americanum* Miller

Currants and gooseberries are introduced species from Asia and Europe. The red currant, for example, was brought into Massachusetts by English settlers in 1639.

Habitat • Depending on the species, wild currants can be found in moist woods or thickets, in rocky places, in fields, on sunny slopes, and in open areas in the woods.

Description • For the most part, currants lack thorns and have smooth-skinned fruit. The exception is the species *Ribes lacustre*, known as bristly black currant or swamp currant, which has prickly or spiny stems. Gooseberries have thorns and bristly fruit.

Wild currants are similar to the cultivated varieties in appearance. They are low, erect shrubs. The leaves have three to seven lobes, and are shaped like maple leaves with toothed edges. The five-petalled flowers are small and greenish-white to reddish-purple in colour, depending on the species. The plants bloom from spring through summer into August.

The flowers are followed by round juicy fruit which range from red or purple to black in colour, again depending on the species. All the species are edible, but the blackcurrant and the wild red currant are tastier than the others. Fruit can be picked from July to September.

Usable parts • berries

Common uses • Currants are rich in pectin and make excellent jellies, pies, and sauces. They may be used fresh, dried, or frozen, whole or pureed. Currants are often used to make excellent wines.

Other uses • Bottled blackcurrant syrup has been an import to Canada from Britain for years, and is readily available on many supermarket shelves. I grew up on it, stirred into boiling water for a winter warm-up. Blackcurrants (*Ribes nigrum*) are the basis for the famous French crème de cassis liqueur.

Medicinal properties • Currants are rich in vitamin C.

In gardens • Currants are easy to grow and even one plant is worth finding room for. They will grow 60–120 cm (2–4 ft) high and need about as much room around each bush. They like the shade and will grow in parts of your garden, such as along a north-facing wall, where other plants don't do well. While currants are heavy feeders, too much fertilizer (well-composted manure is best) will produce lush greenery and little fruit. About two inches of mulch, made from the same composted manure, may be left on year-round. Mulch feeds the plants, keeps the plants from drying out and prevents weeds. Be careful digging in your mulch as currants' roots grow shallow.

Dandelion

Taraxacum species • Aster family (formerly the Composite family)

Other names • Lion's tooth, wild endive, priest's crown, fortuneteller, blowball, *pissenlit*

The oldest of many names that have been applied to dandelion is the medieval Latin name, *dens leonis,* meaning lion's tooth; this became *dent-de-lion* in France. The French also called it *pissenlit,* or wet-the-bed, because of its diuretic effects.

The genus name *Taraxacum* derives from the Greek *taraxos* (meaning disorder) and *akos* (meaning remedy) as a reference to the curative action of the plant recognized by early healers.

Where found • Originally a native of Europe, dandelion arrived in North America with the first European settlers, perhaps the Acadians, and now has become extremely commonplace. Many species are found worldwide; the larger species are the ones most frequently used as food. In the Maritimes, *Taraxacum officinale* Weber is the most widespread and the most familiar.

Habitat • Dandelions grow in pastures, on waste ground, along roadsides, and especially, it seems, on lawns, where they are usually unwelcome. Once established they are a hardy perennial that is very difficult to get rid of. In parts of Europe, however, dandelion is widely cultivated in home gardens and commercially for its leaves and flower heads. A mainstay in Italian cooking, bunches of dandelion greens are sold in markets in many North American cities.

Description • The familiar bright yellow, single-flowering heads grow on hollow stems that shoot straight up from the root. The flower heads are sensitive to light and weather, and close up when darkness or storms approach. These turn into seed heads that drift on the wind like tiny parachutes, becoming one of nature's most efficient seed-dispersal systems. Each flower head contains about 180 seeds, each with its own tiny parachute. The long shiny leaves have jagged edges. It is from these that the plant likely got its name—*dent-de-lion.* The leaf shapes, however, vary considerably on a single plant and between plants. In fact, this is one of dandelion's "survivor" strategies—in dense growth it can raise and elongate its leaves above the competition. The grooved leaves hug the ground in a rosette shape so that as rain falls on them, it is carried straight to the centre of the rosette and directly to the root. The rosette leaves are so densely packed they act as a sort of umbrella that otherwise would prevent the rain from reaching the root.

The thick taproot, single or branched, can grow up to 30 cm (12 in) long in rich soil, and sprouts a number of flower stalks. Roots gathered in October have stored up their winter food reserves in the form of inulin (a kind of sugar that replaces starch), giving the roots a heavy, firm feel. In spring, the food reserves have been used up by the plant so the roots will be shrivelled-looking, with minimal nutritional and medicinal properties. (Inulin is soluble in hot water so boiling the roots extracts it.)

Dandelion stems contain a milky, bitter-tasting juice. When this comes in contact with the skin, it turns brown and is sometimes hard to remove.

Usable parts • entire plant: roots, leaves, flower heads, flowers

Common uses • The dandelion has unexpected virtues. The entire plant is edible, making it an important wild food. It can be collected quickly and easily and has a long season of availability. Be sure to pick dandelions from a clean area, not a sprayed site or a dusty roadway. By cutting off the cluster of leaves with the crown of the root, the leaves will hold together, making it easier to wash the plant.

The roots should be dug up with a shovel or a digging tool in the fall to obtain the most stored nutrients, though they can be gathered at any time of the year. Be careful to get under the whole root or it will break and produce more plants from the broken bits. Select the large fleshy roots for roasting. It is easier to collect roots from good rather than stony soil. One way to clean the roots is to shake off as much earth as you can and then leave them in a container in a running stream, or shake a handful at a time under water.

In Europe, dandelion leaves are used as a potherb and salad green; the roots are used as a coffee substitute. The petals are used to make a clear, golden wine (remember Ray Bradbury's long ago book, *Dandelion Wine*?). The dried leaves were made into herbal beers in parts of England, and this use would have been brought to Canada by the early migrants.

At least one Canadian seed catalogue lists dandelion as a boiling green and recommends tying up the plant to improve the flavour of the heart. The leaves can be eaten raw or cooked, and are best early in the spring, when they are tender; later they become bitter, but can still be used. If you dislike the bitterness, boil them, changing the water twice. Dandelions also lose their bitterness in the fall after being frozen several times and so can again be eaten raw. Young leaves are excellent raw in salads, alone, or mixed with other greens such as spinach or lettuce. Or use them as a sandwich filling as you would watercress, adding salt, pepper, lemon juice, or mayonnaise. As with other leaf vegetables, tear the leaves instead of cutting them. For a novelty vegetable, cover some dandelion plants with a piece of dark plastic, a bucket, or an upside-down flowerpot. The leaves will continue to grow but will become whitened and exotic looking, like an expensive endive.

Cooked, the greens can be used like spinach (although they take a little longer to cook than spinach), seasoned with garlic, nutmeg, grated onion, lemon juice, or its peel. Stir fry the whitish hearts of young plants (found at the base of the leaves) in butter and fold them into an omelette. Or wilt them in leftover hot bacon grease and a little water and cook for four or five minutes.

An old Welsh recipe mixes grated or chopped dandelion roots (use two-year-old roots) with the leaves in a salad. Two-year-old plants produce the large, well-formed roots that can also be used in cooking; the younger roots are too slender. A caffeine-free coffee substitute or additive can be made from the roots of dandelion or chicory roots or a combination of the two. Air-dry the roots, which will take about two weeks, or roast them in an oven at a low temperature for about two hours until they are dry and brittle enough to snap into pieces. Then put them through a coffee grinder and store them in an airtight tin or jar, making sure no mould appears. Roots can also be cleaned and cooked like carrots until tender, eaten alone or with a dressing, or added to soups and stews. As with the leaves, you may prefer to boil them in two changes of water.

You can make a dandelion tea by boiling the washed roots for about half an hour, then adding your favourite herbs and sweetener. A tea can also be made by steeping dandelion leaves in boiling water. Before they grow above the leaf bases, the young flower buds can be added to salads or boiled as greens.

Dandelion wine is made by fermenting the blossoms, avoiding any bits of the stem or the green bracts around the flower heads. Old herbals say it is extremely good for the blood! The petals can also be boiled in water with sugar, pectin, and orange peel to make a honey-like conserve.

Other uses • Dandelion is so commonplace it is easy to bypass its aesthetic qualities. The yellow flower heads make a beautiful dye for wool. A closely related species to *T. officinale*, *T. platycarpum*, was used in Japan by florists for its decorative properties and treasured by the Dandelion Society of Japan. In 1860, some 214 varieties of dandelion were grown in Japan in shades of copper, white, orange, and black.

Children love dandelions for necklaces, bouquets carried home to mother, and for telling time by puffing on the seed heads until there are no white fluffy seeds left to count. Dandelion stems turn into little snakes when they are dropped into water—cut them into pieces and split them at the ends. A good project for a summer's day would be counting the individual seeds in a dandelion seed head to see how many there are.

Medicinal properties • Dandelion has been used for centuries as a folk medicine and is also a component of many patent medicines. As far back as the thirteenth century, dandelion was alluded to in Welsh medicines. Even earlier, it was first mentioned by Arabian physicians of the tenth and eleventh centuries as a sort of wild endive under the name *Taraxacon*.

Herbals have long recognized dandelion as a system cleanser and detoxifier because of its bitter juices. As it is not poisonous, it can be taken in large doses. It stimulates digestion, improves the liver function, and cleanses the kidneys. It was also a traditional treatment for scurvy.

The greens are rich in iron, potassium, calcium, and phosphorus. They contain vitamins B and C, linolenic acid, and more vitamin A than any garden plant, and are one of the best-known plant sources of copper. The blossoms are rich in choline, one of the B-complex vitamins.

Dandelion root is recommended for medicinal use as a very mild laxative, and the leaf as a mild diuretic according to scientific studies carried out in the 1930s. The German Commission E suggests that it may have appetite stimulation and diuretic properties. For diuretic use, make a tea from the young leaves. Herbalists have recommended that its milky sap be wiped frequently on warts.

In gardens and on farms • Dandelions like gardens and thrive in them. Grass is no match for such opportunistic "weeds," nor is any other plant. Dandelion covers the ground with its rosettes to prevent other seeds from establishing and they also give off ethylene gas that retards the growth of nearby plants and reduces competition. If you have too many dandelions, however, your soil is probably depleted from water runoff, use of chemical pesticides or fertilizers, compacting of topsoil from overuse, or nutrient-poor topsoil. Dandelion's long taproots are designed to draw moisture and nutrients up to the surface from depths in the earth that other plants can't reach. Use a special gardening tool to dig them up. One designed specially for dandelions will have a long stem with a two-pronged end to grasp the root. Twist it as you pull it up, being careful to bring it out in one piece. Remember that any pieces left in the ground will make more dandelions. If you have more than you can handle, you should consider improving the quality of your soil organically—and think about growing something other than grass—dandelions, for example!

In the garden and in the wild, dandelion sustains many insects, from bees to small flies, moths, and butterflies, as well as small birds (who like the seeds) and animals such as squirrels and white-tailed deer. Because it blooms so early in spring, dandelion produces pollen and nectar for bees before other sources are available. And because it flowers well into the autumn, it is a source of honey after other flowers have completed their cycle.

Dandelion is also an excellent compost plant, especially for dressing shallow rooted plants that can thus take advantage of the nutrients dandelion draws up from deep in the ground. If you're going to grow dandelion as a crop in any quantity, pick off the flower heads as soon as they appear so that you don't seed all of your neighbours' land too. Harvest the roots from two-year old plants in October.

On farms, it is an advantage to have dandelions in the meadows as their consumption increases the flow of milk in dairy cattle. It is a valuable food for rabbits, especially in the spring, and pigs and goats also like dandelion (though sheep and horses don't). Rotating the crops your animals graze on is a nature-based method that keeps dandelion on your side.

Day Lily

Hemerocallis fulva L. • Lily family

Other names • tawny day-lily, orange lily

Where found • Day lily is fairly common and widespread throughout the mainlands of the region, but especially in the Annapolis Valley of Nova Scotia and westward into Ontario.

Habitat • A Eurasian genus that escaped from gardens, it is now naturalized in fields, wastelands, and along roadsides. It also remains a popular garden ornamental.

Description • Day lily's picturesque, bright orange flowers catch our attention as we drive along country roads; they will quickly populate a roadside ditch. As perennials, if left on their own, they become abundant and can thrive just about anywhere, multiplying and spreading quickly. They are easy to transplant and require very little maintenance, making them a favourite of gardeners.

In appearance, the day lily resembles the familiar lily. It may grow up to 185 cm (6 ft) tall. The unspotted blossoms, 6–8 cm (2–3 in) long, face upward from the top of a leafless flower stalk. Its long, swordlike leaves are a beautiful shade of light green. Its root is a tangle of small, elongated tubers. Day lily flowers from June through August.

Usable parts • young shoots, flower buds, flowers, tubers

Common uses • Day lily is a beautiful cut flower, but it lasts only a day; hence, its name. In eastern Asia, its uses as a food plant are better known than they are in North America. It actually has quite an interesting range of uses.

The early spring shoots may be used like asparagus, or you can add them to salads. The young flowerbuds can be used like green beans, eaten raw or lightly steamed. Add them to omelettes or stir fries, or they can be pickled like capers. The fully opened flowers look beautiful on top of a mesclun salad. Both the buds and the flowers can be battered and turned into fritters. Use them to flavour homemade chocolate sweets such as mousse or truffles. Older, wilting flowers that are a little past their prime can be added to stews or soups as an aromatic thickener, and they can be dried or frozen for later use.

Carefully picked from the centre of the flower, the long filaments that extend upwards can be used like saffron as a flavouring for sweets, rice dishes, and as a sprinkler on vegetables. Use them fresh or dry and store them carefully for later use. The spring tubers can be added to salads but choose crisp, snow-white ones. Tubers dug at any time of the year can be boiled or roasted like corn.

Dock

Rumex L. species • Buckwheat family

Other names • The various docks are also called Indian rhubarb, sorrel, common sorrel, meadow sorrel, sourgrass, and wild rhubarb. Various species may be used similarly. Among the more common species are:

Rumex acetosa—Sour dock
Rumex crispus—Curled dock
Rumex obtusifolius—Common dock

Dock is one of a number of plants called wild rhubarb (as is mountain sorrel, which is part of the same family but not included in this book). *Rumex crispus* is the dock most commonly included in books on edible wild plants. It can be identified by its oval leaves, whose edges tend to curl towards its midrib as they mature (*crispus* means curled).

Where found • Docks are among the most common and widespread of the region's wild vegetables. The various species adapt to such a wide range of environments that you should have little difficulty in hunting down one or another of them.

Habitat • Various species of dock are found in gardens, fields, waste places, and along roadsides; others grow in moist meadows or along riverbanks and lakeshores.

Description • Wild docks somewhat resemble cultivated rhubarb, especially rhubarb that has gone to seed. The docks are taller than rhubarb, though, and their leaves are generally smaller. Dock leaves are green, sometimes with a red tint, and are larger at the base of the plant, decreasing in size upwards. In shape, they are tapering or rounded; some are arrow-shaped. The plants can often be identified by the sour taste of these leaves. There is a papery membrane where the leaf stalks attach to the stem. A long, wand-like cluster of tiny, three-angled flowers, green to purple in colour, is followed by the fruits. The fruits are enveloped in thin-walled "wings." In the fall, these fruits become rusty-brown in colour.

Usable parts • leaves, seeds

Common uses • Gather the greens early in spring or late in the fall, but not in the summer, as warm weather turns them bitter. You can harvest many times from the same plant by leaving the plant itself intact and breaking off only young, tender leaves. The leaves and stalks make a good potherb or cooked vegetable. For best results, boil them in two changes of water to remove the strong taste, using very little water in the second boiling so they don't become watery. Use them in recipes that call for cooked spinach, or in place of rhubarb in a pie. Many people find dock leaves too tough to be used raw in salads.

The seeds are plentiful and can be easily gathered. If you dry them in the oven at very low heat, and then grind them in a blender or by hand, the resulting meal can be used as a flour additive in breads or pancakes. Dock is related to cultivated buckwheat, which has many such applications.

In medieval England, the leaves of common dock (*Rumex obtusifolius*) were used to wrap butter, hence another common name, butter-dock.

Medicinal properties • In medieval times, the leaves were used to treat skin problems, respiratory problems, and liver disorders. The roots were used as a laxative. Because dock and nettle often grow together, if you suffer from nettle stings, look nearby to see if you can find a dock. Rub the fresh leaves on your skin for relief.

Caution • *Rumex* species may cause a skin irritation depending on the degree of contact and the susceptibility of the individual. Most people have no cause to worry. Dock leaves are edible, but they should not be eaten in excessive quantities—they do contain oxalic acid, which is safe in regular amounts. Wild dock is often called wild rhubarb and is, in fact, a member of the same family as cultivated rhubarb; however, rhubarb leaves are quite poisonous. The roots of dock should not be eaten as they contain high concentrations of tannic acid.

Evening Primrose

Oenthera biennis L. • Evening Primrose family

Other names • Varieties are called sundrops, small-flowered evening primrose, and yellow evening primrose. The name describes how the flowers open in the lower light conditions of morning and evening. Under bright sun and dry conditions, they will close up or wilt.

Where found • Eighty species of evening primrose are found in the New World with four growing in the region. The one I am most familiar with is *O. biennis*, but you could use any of the local varieties in the same way. From North America, evening primrose was introduced into Europe and Asia as a vegetable. The variety *O. grandiflora* is an escapee from cultivation. It is distinguished from the other species only by the larger size of its petals. It ranges from British Columbia to Newfoundland and into the United States as far south as Florida.

Habitat • dry sandy soil, gravelly shores, forest edges, roadsides, old gardens, ditches

Description • Evening primrose is a biennial. In its first year, a broad rosette spreads out close to the

ground. The following year, a hairy stem develops out of the rosette. The leaves are narrow and alternate on coarse, erect stems that grow up to 120 cm (4 ft) tall or more. They have slightly toothed margins. The taproot is fleshy and branching.

The beautiful, showy flowers are a soft shade of yellow. In the morning, blossoms that opened the evening before will be fresh-looking and will remain so during the day. In hot and dry conditions they will shrivel and dry up. The flowers have four petals that range in shape from narrow to oval or even heart-shaped and have eight stamens. Flowers of various species bloom between June-July into September-October, and are pollinated by moths. *O. biennis* has fragrant, showy flowers with wide petals. The elongated seed pods are green in colour and are covered in dense surface hairs. The pods split open into four sections, each packed with tiny reddish-brown seeds. After the seeds have matured, the plant dies, as do other biennials.

Usable parts • leaves, roots, flowers

Common uses • The basal leaves can be cooked like spinach and added to soups and stews. The roots can be boiled like parsnips or carrots, but only use the roots of first-year plants (before they flower in their second year). Boil the roots in several changes of water and peel them before serving. The petals have a delicate, slightly bitter but pleasant taste. They make a beautiful addition to salads but they don't store well, even overnight. Lay them on top of your salad or alongside the other ingredients—they bruise easily and turn mushy.

Medicinal properties • Evening primrose is an excellent source of vitamin E. The extract may be purchased as an oil in capsule or liquid form. It contains gamma-linolenic acid (GLA), an essential fatty acid. Often lacking in the western diet, GLA supports the body's hormone balance. When GLA is deficient, the effects of diabetes, premenstrual syndrome, and other disorders are more pronounced.

Herbalists recommend evening primrose supplements for brittle hair, dry eyes, arthritis, and other conditions. Too large a dose over time may cause minor side effects such as skin rash or nausea but this would be unusual. Evening primrose was once called the king's cure-all because of its cold-fighting properties. To make a cold remedy, gather the leaves and stems and steep them in boiling water. You can also dry them for winter use. In Britain, an ointment was made from the leaves to treat eczema and other skin problems.

In gardens • Plant *O. biennis* for its culinary uses and for its perfume. Evening primrose is a prolific seeder, but watch for when the pods are just about ready to open if you plan to gather the seeds. I tie a ribbon on the stems because once the flowers are gone, the plant is harder to distinguish among all the other plants and grasses. I scatter the seeds in my garden amongst the other wildflowers. It is a biennial but they should flower in the first year if you plant early and in good soil.

Fiddlehead

Matteucia struthiopteris (L.) Todaro • Wood Fern family

Other names • ostrich fern

The Latin *struthiopteris* means "resembling an ostrich." The name comes from the shape of the mature fertile frond (the leaf of the fern), which resembles an upright ostrich feather. The "fiddleheads" are the coiled crosiers (the curled tips).

Where found • Fiddlehead is widespread from Nova Scotia to British Columbia and into parts of the United States, and is found almost worldwide in the northern temperate zone.

Habitat • low-lying wooded areas, especially along streams

Description • No book on edible plants of the Northeast would be complete without mention of this fern. Nearly all ferns have fiddleheads but the young, unfurled leaves of the ostrich fern (*Matteucia struthiopteris*) are unlike any other. They push their way up through the ground during April and May and should be picked as soon as they reach an inch or two in height. A specialty of the Maritimes, they have found their way onto the menus of upscale restaurants across North America.

Usable parts • the unfurled heads or croziers

Common uses • To prepare them for cooking, brush them carefully to remove the brown scales. Boil them in a small amount of lightly salted water for about ten minutes or steam them for twenty. Don't overcook them or they will lose their delicate flavour; on the other hand you should make sure they are well-cooked for safety's sake. They can be eaten hot with butter and freshly ground pepper (I like mine on toast) or served cold in a vinaigrette. Because they have a short season, many people like to bottle their fiddleheads in brine, or freeze them by blanching them for a couple of minutes, and then packing them in airtight containers. I have bought them commercially both frozen and bottled.

In gardens • The ostrich fern is easily grown in a home garden as a crop among your other vegetables or as a landscaping feature (sometimes it will establish itself on its own). Ferns produce what seem like trillions of spores (they will be like a cloud in the forest during the summer), but they don't usually produce new plants this way. More commonly, new roots are produced annually from the base of that year's fronds.

Choose an area of your garden that has partial shade. Ostrich fern flourishes in a variety of soils but does best in well-drained soils rich in organic matter. Prepare the bed well before planting by digging in well-composted organic matter (A pH range of 5.5 to 6.5 is recommended). Choose a place where the ostrich fern is abundant and healthy looking and do not remove more than one in seven plants. Dig the crowns when they are dormant in the fall (after the first frost in September or October) or spring (just before the crosiers emerge), selecting only young crowns. You may plant the crowns in rows or beds but I try to space them as they grow in the wild. Set the crowns so that the soil covers just the top of the dormant crown, which will encourage new root growth, and cover with organic mulch such as decomposed sawdust, peat moss, or leaf compost to about one inch.

Be careful to keep weeds away until the plants become well-established, after which the ferns will have spread out far enough to make their own weed-shaping canopy. You should wait two to three years before you begin harvesting the fiddleheads, and do not harvest more than half of the fiddleheads from any one crown. Each spring, enrich the bed with organic fertilizer, as nature would do in the wild.

Fireweed

Epilobium angustifolium L. • Evening Primrose family

Other Names • willow-herb, large willow-herb, rose-bay

The scientific name *angustifolium* means "narrow-leaved." *Epilobium spicatum* is an out-of-use synonym.

Where found • As its name implies, fireweed is one of the first plants to spring up after a forest fire, logging, or avalanche. It helps to restabilize the newly broken ground by conditioning the soil, facilitating plant succession by preparing the ground for a more diverse plant community.

A British species, it is found more commonly in Scotland than in England. Now wide-ranging and common in North America, it extends north of the treeline from Greenland to Alaska, south into British Columbia and California on the west coast, and down into North Carolina on the east coast. *Epilobium angustifolium* is the floral emblem of Canada's Yukon territory.

Habitat • The beautiful, showy, purple-pink flowers of fireweed are a breathtaking sight along streams, on damp slopes and river bars, and beside lakes. (Occasionally pure white flowers will be seen.) It can often be found covering very large areas in meadows or along roadsides, flourishing in bright sun and bare mineral soils. In fact, it can take hold just about anywhere except in deep shade. On Heron Island in northern New Brunswick, I walked through a forest of fireweed growing so tall I couldn't see over the top of it. Here it had taken over a clearing amongst the trees on an abandoned farm site. On the other hand, I've seen a single plant growing in the middle of a city in a back alley or alongside a garage.

Description • The leafy stems grow from about 60 cm to 2 m (2–7 ft) tall. The leaves are pointed and narrow like willow leaves, hence fireweed's other name, "willow-herb." These grow out and upwards from the main stalk for its full length and are spirally arranged on the stem. The top sides of the leaves are a darker green that the undersides. Shoots appear in spring; leaves fill out in the summer. The stalks will develop out of a rosette.

The four-petalled flowers bloom first on the lower part of the stem, followed by those higher up on the stem. Thus, one stem will stay in bloom for several weeks and seed pods begin to mature on the stem while fireweed is still in flower. The male bloom, which is pollen-bearing, evolves into a female flower that is rich in nectar. In the absence of pollinating insects, the female flower is able to open up to self-fertilize.

The flower buds droop while the seed pods angle upwards. These long, narrow seed pods split open when ripe and their fluff, sometimes mistaken for dandelion seeds on the breeze, is carried off for many miles through the air. The root is perennial and very brittle. Take care if you are trying to remove it from your garden as, like dandelion, every little piece of a root will grow into another plant.

Usable parts • young leaves, stalks, shoots, flower buds

Common uses • The young leaves can be used dried or fresh to make a tea on their own, or added to other mixtures. Boil the leaves well. Young, tender shoots are good raw in salads or cooked as a vegetable. Try boiling or steaming the leaves the way you would spinach, and serve with salt and pepper and vinegar or lemon juice. The older leaves and stalks can also be eaten, but the stalks should be peeled first, to keep them from being bitter and tough, and they should be boiled in two changes of water. The flower buds can be added to fruit or vegetable salads for colour or just nibbled fresh off the plant as you pass by.

Sir John Franklin (1823) noted that the young leaves of *Epilobium angustifolium*, under the name of *L'Herbe Fret*, were used by Canadian voyageurs as a potherb.

Other uses • First Nations made twine from the fibrous outer covering of the stems.

Medicinal properties • Fireweed shoots are rich in vitamins C and A. Old herbals prescribed all the *Epilobiums* similarly for stomach and urinary disorders, whooping cough, and asthma, and for migraines.

In gardens • Fast-growing fireweed can become a showy garden perennial that is easy maintenance in a garden or wildflower meadow—it won't need watering or mulching. You will need to keep an eye on it, though, so it doesn't become invasive and dominate less hardy plants. It will likely just show up in your garden one day or you could transplant it via a section of the root crown or rhizome. You may also collect the seeds in the fall and scatter them where you want it to grow.

Butterflies, hummingbirds, and bees sip at the flower's nectar—even a few plants will attract them to your garden. I recall being chastised by an animated female hummingbird in my very own garden, presumably because I was too close to a fireweed plant she had claimed for her own!

Fireweed produces excellent honey that is dark in colour and rich-tasting. I know one beekeeper who sets up his hives anywhere there has been a recent clearcut.

Wild Geranium

Geranium robertianum L. • Geranium family

Other names • herb-robert, bloodwort, fox geranium, crane's bill, meadow-geranium, dove's foot, felonwort

The genus name may derive from the Greek *geranos*, for "crane," in reference to the longish fruit/seed pod which resembles the beak of a crane.

The origins of the name herb-robert are obscure. While the name derives from the medieval Latin *herba Roberti* , the particular Robert it refers to is lost in history. Some say it honours St. Robert, who founded the Cistercian Order; he was born on April 29, about the time the plant usually flowers. Or it may be a corruption of the very old name rubwort in British plant lore. It has even been suggested that the name commemorates Robin Hood. Or perhaps it derives from the Latin *rubor*, meaning red, the colour of its flowers.

Where found • The Geranium family is known for its many ornamental cultivars, of which there are about three hundred species worldwide with pink, purple, or white flowers. *G. robertianum* ranges from British Columbia to Newfoundland and is also found in Africa and Eurasia. It was introduced into North America from Europe.

Habitat • gravelly areas, shores, driveways, cleared areas, rocky slopes, forest edges

Description • The wild geranium's looks are deceiving. It is a delicate-looking plant, but it is a hardy and competitive annual. It may spread along the ground or grow up to 60 cm (2 ft) tall. Its reddish stems are branching, sticky, and hairy. Most people dislike the plant's scent when it is handled.

The leaves are lobed and finely dissected. Those on the stem are usually opposite. They are soft green in colour with a purplish edging. The tiny, delicate flowers are five-part with distinct petals and are pinkish in colour, and each holds ten stamens. The species has a long flowering season between May and October.

When the plant matures, the "beak" separates from the column and coils outward. Then the seeds are ejected by a sudden opening of the pods.

Common uses • I've included it here as a reminder of how ancient some of the plants that we take for granted are, and how they have importance in our history for their own sake.

Medicinal properties • St. Rupert of Salzburg (who died approximately AD 718) was known for his ability to treat a painful skin disease, erysipelas, using wild geranium. Some say he discovered the plant's ability to stem bleeding, thus its name, bloodwort. It was also named felonwort because it could treat felons, which are inflammations near the toenails or fingernails. Today, it is known primarily as a treatment for bruises and skin irritations, applied externally, but we don't use it very much anymore. Its astringent properties come from its tannin content.

In gardens • Wild geranium will just appear in your garden and your planter boxes if you allow it. I always leave some growing because of its history and because of the prettiness of its tiny flowers. If you do want to remove it, wait until it is 5–10 cm (2–3 in) high so you can grasp it and pull it up more easily.

Goat's Beard

Tragopogon L. species • Aster family (formerly named Composite family)

Other Names • oyster-plant, salsify

There is also a true salsify, a cultivated plant, which appears in seed catalogues. The roots of wild goat's beard are smaller than those of cultivated salsify, but taste just as good.

Where found • Goat's beard, originally a Eurasian species, was introduced to North America from Europe. It has become nearly as common as its relatives in the same family, dandelion and chicory. The Aster family is one of the world's largest families of flowering plants.

Habitat • roadsides, fields, waste ground, and lawns

Description • Goat's beard flowers and roots look very much like those of the dandelion; however, the leaves of the two are very different. Dandelion leaves are jagged, whereas goat's beard leaves are more slender and grass-like. Goat's beard leaves alternate on the stem. The stems, leaves, and roots contain a bitter milky juice.

Goat's beard stands erect on hollow stems, growing up to 1 m (3 1/4 ft) high. The flower heads are pale yellow, and they unfold in the morning and close up at midday. Collect them from June to October before they turn to fluff—the tiny "parachutes" that scatter the seeds on the wind. The leaves are best in early spring; the roots in spring or fall.

Usable parts • roots, young leaves

Common uses • The young greens can be used raw in salads or cooked as a vegetable. They should be boiled for about five minutes. The roots, which taste somewhat like oysters, should be cleaned well. Peel and slice them, and boil immediately to avoid discolouration. Cook for about half an hour, and serve with salt, pepper, and butter, or add to soups or stews. (You can also peel the roots after boiling, if you wish.) The roots can also be roasted and ground into a coffee substitute. The milky juice, partly hardened, can be chewed like gum.

Medicinal properties • Herbals recommend using goat's beard roots in a tea as a remedy for heartburn, digestion problems, and lack of appetite.

Goldenrod

Solidago species • Aster family (previously known as the Composite family)

Where found • Goldenrod is widespread and common; a hundred species of the genus are native to North America, and many of these are found in the Atlantic region.

Habitat • Depending on the species, the habitat ranges from open woods to riverbanks and lakeshores. Some prefer dry habitats; others damp ravines.

Description • Plants range in size from low to tall, and their descriptions vary among the species. All have thin stalks and slender leaves without lobes or teeth. The leaves exude a scent like anise when crushed. The flowers, which appear from July to September, are tiny, yellow, and grow in clusters. From a distance they look like clumps of fuzz.

Usable parts • young leaves, fully opened flowers

Common uses • Gather leaves and use them fresh, dry them in the shade for later use, or store them in airtight jars. A dozen per cup, steeped in boiling water, make a mild, clear yellow tea, which many herbals recommend as a stimulant or to dispel gas. The leaves also make a useful potherb.

In the United States, around the time of the Revolution, the flowers of wild goldenrod were picked, dried, and shipped to China for use as tea.

Other uses • Goldenrod flowers produce a strong yellow dye. On my first archaeological dig, I remember being shown how First Nations varied the dye colour by soaking the flowers in a rusty bucket.

Caution • People who suffer from hay fever may react to goldenrod. They should use it cautiously.

Medicinal properties • The genus name comes from the Latin *solidago*, meaning "to make whole," in reference to the plant's healing properties.

In gardens • Most people prefer to pull goldenrod from their gardens, but it can become part of a diverse ecosystem if you keep it under control and enjoy it for its bright colour and its food and medicinal uses.

The leaf print shown here is of Canada goldenrod, which has denser leaf growth than other varieties.

Gold Thread

Coptis trifolia (L.) Salisb. • Buttercup family

Other Names • mouthroot, cankerroot, yellowroot, tissavoyanne jaune, fil-d'or

The French-Canadian name, savoyan(e), comes from the Mi'kmaq word, *tisavoyane*, meaning "dye for skins."

Where found • From Labrador to Alaska and south into the U.S.; also in Eurasia.

Habitat • coniferous forests, roadside banks, swampy and damp areas, mossy woods, on hummocks at the forest edge

Description • Gold thread's leaves, which at first glance resemble strawberry leaves, consist of three oval-shaped leaflets (widening slightly towards the tip), toothed and rounded on the lower side. The leaves are evergreen, thus you can find gold thread under the snow throughout the winter. The leaves grow up and out on long stalks from the roots that are the plant's distinctive feature and the source of its common name. The roots grow in a tangle of bright yellow threads and can be found by scraping back the forest debris and digging down an inch or two. They can be pulled up by the handful in areas where they are abundant.

The flowering stalk extends beyond the leaves and holds a single white star-shaped flower with five to seven petals; it blooms from mid-May to the beginning of June and is followed by three to seven long beaked pods.

Usable parts • roots

Common uses • The French learned to use the roots for dyeing skins and wool from First Nations, who also used it to dye feathers and porcupine and bird quills.

In the early 1800s, gold thread was sold in druggist shops in Boston. By 1879, it was selling for seven cents a pound, primarily for use as a medicine for sore eyes and sore mouth. By the turn of the century, the price had risen to one dollar a pound.

Medicinal properties • In 1868, gold thread was included in the *Canadian Pharmacists' Journal* in a listing of Canadian medicinal plants. From 1820–1882, it was listed in the United States Pharmacopeia. It contains a mild sedative, the alkaloid berberine, to which its medicinal uses are attributed. You should avoid berberine if you are pregnant. In extremely large doses, berberine may cause problems and it is probably most effective and safest if taken for short periods of time—no more than a few weeks.

The taste of gold thread roots is very bitter and astringent. Traditionally the roots were used to treat soreness in the mouth, thus one of its common names, cankerroot. It was also used as a gargle for sore throats.

To treat inflammation of the mucous membranes of the eyes, the roots were boiled in water to make a tea. A piece of clean cloth was then soaked in the tea and applied to the eyes. For a child with a sore mouth, gold thread was used as a topical anaesthetic and spread on the gums of the child, usually by placing a cloth saturated with the tea in the mouth. The roots were either chewed fresh or boiled into a tea that was then used as a mouthwash.

In Gardens • If you have a wooded area or a wet, boggy area in your garden, gold thread can be cultivated as a ground cover. It is fairly easy to transplant some of the roots, but take them from an area where the plant is abundant and cover up any digging so the exposed roots don't dry out and the plants you leave behind can continue to flourish.

Gooseberry

Ribes hirtellum Michx. • Gooseberry family

Other names • Common gooseberry, American gooseberry; other varieties of gooseberry worldwide include English and European gooseberry

In 1880s Europe, over seven hundred gooseberry cultivars were available. Gooseberry includes many varieties, some hybrids and some from its hybridization with English varieties. *Ribes hirtellum* is native. The genus *Ribes* includes wild gooseberry and wild currants. Gooseberries are spiny (the Latin name *hirtellum* means "bristly"), whereas currants, with the exception of swamp currant, are not.

Where found • Gooseberries are not uncommon in the region although they are not appreciated as much here as they are in Europe. Their range in North America extends across Canada from Newfoundland to British Columbia.

Habitat • rocky hillsides, along the edges and openings in woods, in fields, roadsides, recent clearings, pastures, and less frequently, in bogs

Gooseberries are often found growing alongside raspberries or rosehips.

Description • In appearance and taste, the wild gooseberry is much the same as the cultivated variety. The fruit is borne on low, sprawling, perennial shrubs, which sometimes form dense thickets up to six feet tall. The hairy leaves, 1 1/4 cm to 13 cm (1/2 in to 5 in) long, are five-lobed. Typically, they have toothed edges and somewhat resemble maple leaves, though gooseberry leaves are more variable in their shape and hair covering. The stems are prickly, with strong, sharp thorns. The small five-petalled flowers are whitish-green to purplish in colour and bloom in June.

The flowers are followed by roundish, smooth, translucent fruit ranging in size from smaller than a pea to the diameter of a dime on a single plant. The berry is a purplish-black colour, growing singly or in small clusters, and contains many seeds. Each berry has a dry, brown tuft at the end that is the remains of the flower. The fruit ripens around September 1, towards the end of the blueberry season, and continues a week or two past the blueberries.

Usable parts • leaves, berries

Common uses • The young, tender leaves can be eaten in salads. The berries can be used fresh or cooked. Gooseberries are popular in England for jams and pies, but in Canada they are rarely seen for sale. My maternal grandmother made the best gooseberry pies! Instead of putting all the sugar in the berries, she sprinkled it on top of the crust so I could enjoy the tartness of the berries by themselves or bite into a mouthful of sweet and sour together.

Though gooseberries are very tart—some say mouth-puckering—even when they are ripe, they are delicious eaten raw with sugar and milk. They are also excellent for preserves, chutneys, vinegars, wines, "fools" (desserts of puréed fruit and cream), sauces, and pies. If you don't like their tartness in pies, try mixing them with apples. Because they are high in citric acid and in pectin, they make excellent jellies. It is usually fairly easy to find a large quantity of gooseberries, since they are often passed over by the more sweet-toothed berry pickers.

You'll need heavy gloves to harvest gooseberries, especially if the growth is dense, because of the thorns on the canes and shoots.

Medicinal properties • Old herbals extol the medicinal value of the leaves. In infusions, they are used for inflammations, fever, and biliousness. Gooseberry contains high concentrations of citric acid.

In gardens • The native gooseberry is winter hardy while the European varieties produce larger fruit; hybrids combine the best of both. European species are more difficult to grow in east coast gardens, but native species are easy to grow in gardens and are less likely to be susceptible to mildew and rust.

Like currants, gooseberries can tolerate partial shade and prefer moist soils. Only one variety is needed for pollination. The best time to transplant is fall or early spring, the earlier the better as the buds begin to open with the first warmer days. Plant them as arbours and living fences, and along buildings or fences in espalier form. The soil should be rich with well-composted manure that can also act as year-round mulch. As with currants, too much fertilizer gives lots of foliage and little fruit.

Wild Grape

Vitis L. species • Grape family

Other names • fox grape

Where found • Wild grape grows only in the northern hemisphere. While it is found growing wild in the region, the Maritimes now has an expanding grape-growing industry, primarily for producing wines. Most of the growers are in the Annapolis Valley.

Habitat • alongside ponds and streams, on roadsides, in open woods

Often found growing up a tree as the tree matures.

Description • A woody vine that produces clusters of purplish-black fruit (up to twenty grapes in a cluster), wild grape is often considered to be a nuisance because it is an aggressive grower and can cover other plants quickly. The vines can climb thirty feet or more using their tendrils to grab hold of the branches and bark of trees and other large plants. The leaves are three-lobed with teeth on the edges. The bark is shreddy and brownish-grey in colour. The flowers, which bloom from May into July, are small and green. The fruits ripen from August through October.

Usable parts • fruit, leaves

Common uses • We know that grapes have been used since prehistoric times to make wine. In fact, it has been said that the history of wine is the history of civilization. While the wild grape was cultivated throughout the world, in North America, where wild grapes are common, it does not appear to have been cultivated pre-contact. The fruit is made into wine, jellies, juice, and pies and is dried into raisins. The seed is made into grapeseed oil and extract.

Grape leaves can be used in cooking—look in Greek and Middle Eastern cookbooks for recipes. You can buy leaves in jars ready for wrapping rice mixtures or you can make your own. Gather large leaves and boil them in salted water for a few minutes until they are pliable. Drain and rinse under cold, running water. Place a heaping tablespoon of your stuffing mixture (try a mixture of cooked rice, raisins, and pine nuts) onto the centre of each leaf near the stem end, then fold the left and right sides over the stuffing and roll the leaf from base to tip. Place the stuffed leaves with the leaf tips on the bottom in a steamer and steam thirty to forty minutes until the leaves are tender.

Other uses • Each fruit contains several seeds that are spread through the droppings of birds and other wildlife. Grapes are an important food source for a variety of wildlife. Raccoons, especially, love them. The bark is also important for constructing birds' nests.

Wild grapes may be difficult for humans to reach in the wild, especially when they are located high in the treetops where birds and animals get to them as soon as they ripen.

Medicinal properties • Grapeseed extract is available as a liquid, tablet, and capsule. Omega Biotech, a company I am associated with in British Columbia, makes a top-quality grapeseed extract that is made into an anti-aging cream as well as capsules to be taken internally. Research has shown that benefits are increased if grapeseed is used both topically and internally.

In gardens • Grow grapes for the fruit, but also to attract wildlife. I've grown beautiful vines from cuttings potted for a year and then transplanted into a planter at the base of my house posts (which they have now wrapped themselves around). I love to watch them change colour with the seasons—from a rich green to deep shades of red.

Highbush Cranberry

Viburnum L. species • Honeysuckle family

Other names • There is some confusion about the common name of this plant. Some people use highbush cranberry and squashberry interchangeably. Others refer to two different *Viburnum* species by these names. Marian Zinck calls *Viburnum opulus* highbush cranberry, whereas *Viburnum edule* is named as mooseberry or cranberry bush. Since my early years foraging in the far north, I have always known *V. edule* as highbush cranberry.

Where found • Highbush cranberry is not related to the true wild cranberries of the Heather family, and you can expect to find them in different habitats.

Habitat • Highbush cranberry is common in moist woodlands and along the shores of lakes and streams. *V. edule* is found in northern Cape Breton, ranging from Labrador and Newfoundland across Canada to Alaska, and south into the United States. *V. opulus* also extends across Canada to Newfoundland in the east.

Description • Highbush cranberry is an upright, deciduous, straggling shrub that grows up to 2 m (7 ft) high. Its bark is dark grey and smooth. The three-lobed leaves are about 4–7 cm (1 1/2–3 in) in length, with coarsely toothed edges. The small, five-petalled white flowers grow in rounded clusters from June to July or August, followed by the fruit.

The berries are large, bright red, and juicy-looking, and grow in bunches. Each contains a large, flattish seed, from which the plant seems to derive one of its common names, squashberry. The fruit is borne in September and October. The berries often stay on the plant during the winter if the birds have not cleared the bushes. Their flavour is improved by the first frost.

Usable parts • berries

Common uses • Although not related to wild cranberries, highbush cranberries resemble them in flavour and are often used in their place. Their only drawback is the seeds, which must be strained out when making juices or jellies.

A delicious juice can be made from crushed and strained berries, and they make an excellent jelly with pectin added to the boiled fruit. Highbush cranberry sauce made with boiled-down berries is good with wild game. Some people find highbush cranberries too tart to eat raw, but I found them delicious and refreshing when I first came upon an abundance of them at the end of a long portage on a hot fall day. Now they are a favourite.

Medicinal properties • Like the cranberry, highbush cranberry is an excellent source of vitamin C.

Hops

Humulus lupulus L. • Hemp family

Other names • common hops, European hops, hop strobile

The name comes from the Anglo-Saxon *hoppan,* meaning "to climb." The species name *lupulus* derives from the Latin *lupus* (wolf) according to Pliny the Elder (AD 23–79) and noted in Mrs. Grieve's *A Modern Herbal* of 1931. It was given the name because it grows amongst willows, strangling them by its light embrace, as a wolf does a sheep.

Where found • In the region are both native and introduced varieties from Eurasia but they are all very similar. Hops range from British Columbia and California to Newfoundland. Some unpromising attempts at commercial cultivation were made in the Annapolis Valley and other parts of the Maritimes. Hops can now be found growing wild up the trunks and along the branches of tall trees on old, untended farm sites; I first saw them in the Campbellton area of New Brunswick and later in the Miramichi region.

Habitat • Able to grow in a wide range of soil types, hops thrive in deep, well-drained non-acid soils containing a lot of loam.

Description • The vines climb clockwise up trees and along fences by the stiff hairs that grow on the stem. The male and female flowers grow on separate plants. The actual "hops" are the cone-shaped scaly seed-vessels of the female plants.

The vine is a perennial rhizome that sends out shoots early in spring. These shoots grow at a great rate, reaching 4 ½ –9 m (15–30 ft) depending on what they can find to cling to. Each year the vines die back to the ground, re-emerging the following spring from their very hardy crowns.

The dark green, heart-shaped leaves have finely toothed edges. They generally, but not always, grow opposite on the stem. The flowers appear in clusters around the end of June or into July. The seed forms around the end of August. Once the hops start producing cones, wait for them to turn light green to yellowish with a dry papery feel. They are then ready for harvest and can be picked, dried, and sold.

Usable parts • stems

Common uses • Hop growing began in the early Middle Ages, with Bohemia and Moravia having long traditions of hop growing. The first documented hop cultivation took place in AD 736 in what is now Germany. Before that they were gathered in the wild, as far back as early Egypt, where hops were used as a medicinal herb rather than as an ingredient in beer. By the end of the fifteenth century, hops began to be used in brewing. This probably started in Flanders and then made its way into Britain, where hops began to replace the traditional ingredients—malted barley and tree barks. By the 1870s, hops were grown widely, including in Wales and Scotland. Introduced into North America from Europe by the Massachusetts Company in 1629, hops production moved from the east coast to the west for its better growing conditions.

Caution • When you're walking through a hop-infested area, be careful not to trip over a strig, the British name for the central stem of a hop plant. Seriously, however, hops are not known to have any contraindications associated with their use medicinally.

Other uses • The tough, fibrous long stems are flexible and very tough. They have been used as a twine and to make paper and cloth. The shoots, which are available in early spring, were used as an emergency food in medieval times. Today in Germany their use has been resuscitated as a gourmet food, served raw in vinaigrette (like asparagus), boiled with fresh herbs, or fried in a batter. On the farm, hops are a valuable food for cows, sheep, goats, and pigs.

Medicinal properties • Ancient Egyptians used hops to treat liver disease and digestive problems. Today we know that acids in the hops produce their distinctive sour taste. They also lend slight antibiotic properties and extend the keeping qualities of beer. According to the German Commission E, hops have been shown to have a mild sedative and sleep-promoting effect, and thus are useful for insomnia and anxiety, especially if taken as a tea before bedtime.

In gardens • Hops can be susceptible to a wide range of problems—frosts, insects, slugs, blight—but you might find them interesting to grow as a hardy perennial in a southern part of your garden where they will get lots of air circulation. Hops are planted in hills and need support of some kind. String twine across a trellis of poles and wire, or construct a simple fencing of polypropylene netting. Old fishnets work well too.

Hops need mulching with organic matter, such as bark or straw, to control weeds and keep in moisture. They can be propagated by cuttings or by purchasing nursery plants but you would do well to seek professional advice—there are generations of it to draw from!

Horsetail

Equisetum L. species • Horsetail family

In this family there is only a single living genus, *Equisetum,* which includes horsetails and scouring rushes.

Other names • scouring rush, devil's gut, mare's tail; the Latin name comes from *equus,* meaning horse, and *seta,* meaning bristle

Where found • Horsetails have spread more widely throughout the world than any other plant. Across Canada and the United States, they range from Labrador to British Columbia, into the far north, and south into the United States.

Habitat • Depending on the species, low fields and embankments, along streams and the edges of ponds, in marshy ground and swamps, from sunny to shady areas. They are prolific and hardy.

Description • A variety of horsetails grow in the region, and all of them are similar in appearance. They have hollow, jointed stems, resembling bamboo a little in this way. The hollow stem provides the nutrients for root growth and spore production. The stems are ribbed and grooved and feel coarse to the touch from the outer layer of cells that contains very hard silicate crystals (thus their use in the outdoors as pot scrubbers). Be careful of your fingers if you use them this way – they'll act like a scouring pad, rubbing off the outer layers of your skin. They will also scratch at your bare legs if you walk through a patch of them.

The minute leaves, which ring each joint, are scale-like and form a circular arrangement. Cones top the green stems, or in some species, the reproductive brown stems. In height, they grow up to 60 cm (24 in), but there are also dwarf varieties. They are annual

or perennial, spreading by spores (rather than seeds) as well as by creeping rhizomes, the underground stems. The spore cases are borne in a cone at the tip of the stem. The rhizomes send up new aerial shoots year after year. The rhizomes store nutrients produced in the previous year for use by the new growth. Each rhizome produces buds that detach easily and form a new plant; thus the horsetail is very successful as a prolific and hardy weed.

Horsetails go back in time almost as long ago as the mosses and liverworts that invaded the land some four hundred million years ago. They predate modern insects, reptiles, birds, and mammals, as well as conifers and flowering plants in the evolutionary process. From fossils, we know that horsetails once grew almost as tall as trees that reached 30 m (98 ft) in height, although these varieties have not survived into the plant world as we know it today.

Usable parts • Shoots of *E. arvense* (field horsetail). For culinary and medicinal purposes, do not use the species *E. palustre* (marsh horsetail), which has been identified as poisonous by some sources.

Field horsetail is common in cultivated fields throughout the northern hemisphere. Its stems are hollow. Marsh horsetail, in comparison, populates swamps and marshes but is rare in the region; only a small portion, less than one-third, of its stem is hollow so it should be easy to identify if you do come across it.

Common uses • The young, green, tender shoots may be peeled and eaten raw or boiled. First Nations in the United States and Canada used horsetails like sandpaper to polish soapstone pipes and implements made of wood and bone, such as canoes and arrows. You can use horsetails to sharpen the edges of scrapers made from mussel or clam shells, or to polish your fingernails. Children might like to try making whistles out of the stems.

First Nations also used horsetail in a decoction as a hair wash. It makes the hair shiny and helps prevent dandruff. Now it is often found as an ingredient in commercial, plant-based shampoos.

Caution • While none of the above uses should cause any problems, it would be wise to remember that the Aleut of Alaska once used an *Equisetum* species to make a concoction that they fed to a hated guest as a magical poison!

Other uses • For emergency use when out on the land, the stem can be sipped from as a source of clean drinking water—the high silica content acts as a filtration system. In the evolutionary process, horsetail found a way to move onto the land by developing specialized cells to transport water and nutrients. The roots may also be used in basketry.

Medicinal properties • The *Equisetums* had a wide range of medicinal properties for First Nations in the United States and Canada. Their stems were used to feed medicine to babies, and the raw stems were given to babies when they were teething. The ashes of several species were mixed together and used as a burn dressing alone or mixed with grease. Horsetail was pounded, mixed with water, and used as a wash externally on parts of the body affected by poison ivy or skin sores. The fluid from the stems was used to wash itchy or sore eyes as an eye medicine. *Equisetum* is known to contain silica, which is essential to life, as well as flavanoids, manganese, potassium, sulphur, magnesium, calcium, iron, and phosphorus. From the soil, horsetail absorbs silica (which is inorganic and chrystalline), then changes it into organic vegetal silica, which can be absorbed by humans.

In the natural wellness industry, horsetail is being extracted into capsules and tablets as a natural nutrient. As a supplement, silica is recommended to enhance calcium uptake as a bone builder and for strengthening the teeth. It stimulates cell metabolism and cell formation and strengthens the immune system. It also helps increase the elasticity of connective tissue and thus is said to help with rheumatic problems, arthritis, and inflammations. You should not dry horsetail to make your own capsules, however, as it can have an abrasive effect on your stomach lining.

Foods known to be extremely rich in silica include unprocessed and unrefined oats and millet. Unrefined barley is also a good source. Soak the raw oats and add them to a breakfast muesli. Dried or fresh horsetail can be added to your bath to ease rheumatic pain.

In gardens • Horsetail indicates a hardpan soil that is poor and acidic. Horsetail, for example, was the first vascular plant to come up through the debris after the Mount St. Helens eruption. If you have too many horsetail plants in your garden, don't try to eliminate them by pulling or digging—their far-reaching roots go too deep and can form a very dense network. The best way to discourage horsetail is by double-digging in a rich compost to enrich the soil and increase the pH level.

I find their foliage quite beautiful and leave areas of my garden for them. They are really maintenance free, and bring a sense of the primeval to the garden. Horsetails are one of our most ancient land plants and have changed very little since they moved, long ago, out of the water and onto the land.

On the farm, horsetail may be used as an occasional fodder plant for fattening geese, or to make the coats of both geese and horses glossy. Horsetail can be a serious problem for horses, however. If cut hay includes twenty percent or more of field horsetail and is fed to horses, especially young colts two to five weeks old, the animals develop symptoms of vitamin B1 deficiency. This is because dried horsetail contains thiaminase, a toxin that destroys the thiamine supplied in a normal diet. It is not a problem for cows, sheep, and goats, because these are ruminants with two stomachs—the toxin is broken down before it causes problems.

Juniper

Juniperus communis L.
Juniperus horizontalis Moench • Cypress family

Other names • *J. communis* is known as ground juniper or common juniper. *J. horizontalis* is known as creeping juniper.

Where found • Two species of juniper grow in the region. Ground juniper and creeping juniper are both hardy evergreens. Ground juniper is the most commonly recognized and the larger of the two; it is common throughout the region. Creeping juniper ranges from Newfoundland to Alaska and south into Maine, but its provenance is more scattered. Ground juniper is a native of Britain, appearing soon after the last Ice Age. It is one of our slowest-growing trees, but I consider it a shrub so it is included in this guide.

Habitat • common and widespread in lowland and alpine areas, stony or gravelly places, on cliffs, and in bogs

Description • One of our slowest growing "trees," juniper grows close to the ground, often forming a thick bushy mat. Its leaves are needle-like, its fruits small and berry-like, bluish-black to purple when ripe (in the second year) or whitish to greenish when unripe. They are called berries here because most people use the term, although they are in fact cones.

Usable parts • berry-like cones, new twigs

The berries can be picked and used all year round, or dried for later use. They are easy to pick in handfuls as they are firm and small.

Common uses • A tea can be made from juniper berries; twigs can also be added. Pick the purple berries, about a spoonful per cup. Crush them and steep in boiling water, covered for five minutes. The crushed berries and tea have a slightly resinous scent, but the taste is mild and not at all bitter. The tea is clear and light amber in colour. Berries also make a good brown dye.

In Scandinavia, juniper berries are commonly used as a spice to add flavour in cooking. Use the berries fresh or dried. You can add a few to any recipe for chicken or turkey stuffing. Try adding three to six berries to a serving of baked beans. The oil from the berries of *J. communis* was used in flavouring gin, known in England as juniper spirit, but the recipe is not given here. Early explorers wrote about infusing the berries in brandy to make a cordial. You could soak some in vodka to get a sense of the taste. The green wood may be burned to smoke meats and fish. When it is dry, however, the wood gives off very little smoke.

Juniper is burned as incense in Siberian and Mongolian households and is used in shamanic ritual. While it may be used in North American ceremonies, most medicine people here use white sage brought in from the United States or they use locally available sweetgrass. As with other smudge plants, the smoke from the smouldering juniper branch or the ground powder is "washed" with the hands over the body to purify and cleanse oneself, other participants in the ceremony, paraphernalia such as drums, and the space in which the ceremony is being conducted. It is also used, as a smudge or simply hung, to clear a home of negative energies or bad spirits. Similarly, in Biblical times in Europe, branches of juniper were burned in the temples as a purifying herb; it was considered a very powerful protection against witches, evil spirits, and devils.

Medicinal properties • Juniper berries are used by herbalists for a wide range of medicinal purposes. They are said to remedy everything from digestive problems and rheumatic pains to joint problems and, in a vapour bath, bronchitis; however, large doses or continuous use may irritate the kidneys and urinary passages.

Roman physicians prescribed it for a variety of ailments too, from pains in the stomach, chest, and side to flatulence, coughs, and colds.

Caution • Some people may be allergic to juniper so try it first in small amounts as a food or medicine. It should not be taken by anyone with kidney problems. Nor should it be used during pregnancy; in fact, *J. communis* was once called "bastard killer" in Britain because of the practice of eating the berries to induce an abortion.

In gardens • Juniper makes an excellent year-round ground cover in areas of the garden where the soil is similar to its preferred habitat in the wild. It mixes well with bearberry and cranberry in an evergreen garden. Because it is slow-growing, it will take some time to establish itself, but once it has spread its branches, juniper will be virtually maintenance-free.

Knotweed

Polygonum L. species • Buckwheat family

Other names • Various species are also called smartweed, bindweed, bistort, knotgrass, matgrass, doorweed, and wild rhubarb. Knotweed is one of several local plants called "wild rhubarb" (as is dock).

The most widely used edible knotweeds are Japanese knotweed (*Polygonum cuspidatum* Sieb. and Zucc.) and giant knotweed (*Polygonum sachalinense* F. W. Schmidt). My focus here (as well as all illustrations) is on Japanese knotweed, also known as wild rhubarb or giant rhubarb.

Where found • A wide range of knotweeds grows in the region and worldwide, the various species preferring different conditions. Introduced from Asia, Japanese knotweed was once cultivated. It is now taking over from less competitive species, including many native plants, and has become a nuisance plant in gardens and in the wild.

Habitat • Knotweeds most often grow on land, but some species are aquatic. Habitats are various according to species, from roadsides and waste places to pond and stream edges and lakeshores. Japanese knotweed prefers waste areas and town edges. Along the roadsides in the region, it grows in large, dense patches that look so lush and established, you would think they were being tended.

Description • The knotweeds get their name from their frequently thickened and swollen joints, which resemble knots, on their stems. The differences among the various species are difficult to describe and not well understood even by botanists. The knotweed species have the following features in common: the leaves alternate on the stem and are unbroken along their edges; the flowers are minute; and the fruits are lens-shaped or triangular.

Japanese knotweed is tall and erect, growing up to several metres in height. Its hollow stems are segmented, like bamboo. The leaves range in shape from oval to heart-shaped. The drooping clusters of flowers that appear in August and September emerge from the leaf axis; that is, where the leaf meets the stem. Japanese knotweed is easily recognized by its size and its thick, segmented stems.

Some knotweed species may be more suitable than others for various common uses. Experiment to select the ones you prefer: peel and taste the stems to see if they have the tart flavour of rhubarb before gathering them in any quantity.

Usable parts • roots, stems, shoots

Common uses • Many knotweeds make good eating as a cooked vegetable. Boil the young shoots and stems for about five minutes, or steam them and serve like spinach. The roots can also be cooked. You may prefer to soak them for several hours in water before boiling. The stems and shoots can be made into a jam or condiment with a little sugar added. Like rhubarb, they make an excellent pie filling, perhaps mixed with strawberries, or a stewed fruit dish to be eaten with custard or ice cream. If the stems are old and tough, peel them before cooking.

Caution • Knotweeds are related to rhubarb. They may, like rhubarb, act as a mild laxative. Also, the leaves of some knotweed plants may cause skin irritation in people who are susceptible, though this is rare.

Medicinal properties • To use Japanese knotweed as a mild laxative, boil it and drink the strained juice. You may want to add a sweetener—try stevia, which has none of the side effects of sugar.

In gardens • Japanese knotweed is very hardy and will be maintenance-free in your garden, but make sure you want it there as, once established, it's almost impossible to get rid of. You may find that it's better to gather it from an established source.

Labrador Tea

Ledum groenlandicum Oeder • Heather family

Other names • continental tea, bog tea, muskeg tea, Hudson's Bay tea, *wish-a-ca-pucca*, marsh tea

Where found • wide-ranging and famous boreal species, often mentioned in historical literature

Habitat • sunny cliffs and ledges, peaty soils, shrubby or moss-and-lichen wasteland, poorly drained pastures, edges of blueberry bogs

Description • Labrador tea is a low shrub with oblong to linear evergreen leaves that range from about 2 1/2–5 cm (1–2 in) in length. The leaves are smooth on top, with the edges rolled back. The undersides are covered with a rust-coloured fuzz. Clusters of tiny white five-petalled flowers bloom in the first part of the summer. The plant's resinous scent, which comes from its oil, gives added enjoyment to a walk in the woods. Remember that Labrador tea grows slowly, so pick leaves rather than whole branches, and harvest from different shrubs.

Labrador tea is one of our best-known wild plants, and was a favourite stretcher or substitute for imported tea among early explorers and traders. It was sometimes used as a table tea.

Usable parts • leaves (these can be picked all year-round or dried for storage), flowers

Common uses • In camp, you can dry the leaves in a cast iron frying pan (to distribute the heat) over a slow fire. Leave them until they become crumbly. Or you can dry the flowers and the leaves carefully in the shade and they will retain their flavour for many years.

The flowers give a more delicate and, to some, a more pleasant taste than the leaves. To make a tea, steep one heaping spoonful of leaves or flowers per person in boiling water for five minutes. The colour is a clear, pale amber-orange. The taste is hard to describe, but pleasant. You could add the dried or fresh leaves to a pot of store-bought tea for a more aromatic drink. For variety add lemon or a sweetener.

Since the leaves can be dug out from under the snow, Labrador tea is an important emergency food resource.

Other uses • The leaves will produce a brown dye.

Caution • Although Labrador tea was recommended in early herbals as a restorative and as generally "good for the system," it does contain a poisonous compound called andromedotoxin that can cause headaches, cramps, paralysis, intestinal problems, or intoxication if taken in excessive amounts. Do not boil the tea or steep it for longer than ten minutes. Other members of the Heather family, such as rhododendrons, also contain this compound.

Medicinal properties • First Nations used it to treat colds and pneumonia as an infusion and as a snuff for nasal inflammation. It was used as a tonic and a diuretic. Burn and scald dressings and poultices were made of the powdered leaves mixed with grease. The Maliseet made an infusion to treat kidney problems, while the Mi'kmaq treated colds with the leaves.

Lamb's Quarters

Chenopodium album L. • Goosefoot family

Other names • pigweed, goosefoot, wild spinach, dirtweed, fat hen, bacon weed

Where found • An introduced weed from Europe naturalized in North America and now common.

Habitat • wastelands, along the banks of streams and near settlements; in gardens, where it is not always welcome

Description • Lamb's quarters is a branching, upright weed that grows to about 1 m (3 ft) in height. Its leaves are triangular or oval in shape, broadly toothed—particularly toward the base—and green on top, pale green underneath. The upper leaves are narrower and toothless. The stems and undersides of the leaves are covered with a whitish mealy coating. There is, however, considerable variation from plant to plant in leaf shape and mealiness.

The flowers are pale green—sometimes turning reddish, particularly in the fall—small, and clustered in dense spikes near the tops of the stems. At maturity each flower is followed by minute, shiny black seeds. There are many similar, mostly edible, species, which are difficult even for a botanist to tell apart.

Usable parts • The young leaves are best, but older leaves can also be used; seeds.

Common uses • Lamb's quarters are considered one of the best-tasting wild edibles. They are very nutritious, being rich in iron, calcium, and vitamins A, B, and C. Boil the leaves for ten to fifteen minutes; like spinach, they reduce greatly in bulk. Or use young leaves as salad greens. The seeds can be eaten raw, but to use them as a cereal, crush and serve with sugar and milk. Or strip them from the stems and boil for about fifteen minutes in salted water. You can also steam them and serve with butter, or add them to soups and stews. The seeds can be dried and ground into a flour for adding to breads or pancakes.

Other uses • Lamb's quarters is a valuable food for poultry; hence one of its common names, fat hen. It is grown in some parts of the world as food for sheep and pigs, and that is where it gets its other names, pigweed and lamb's quarters.

Lupin

Lupinus nootkatensis Donn (lupin)
Lupinus polyphyllus Lindl. (garden lupin) • Bean or Pea family

Other names • The name comes from the Latin *lupus*, meaning wolf, because people believed, mistakenly, that lupin ruined the fertility of the soils it grew in, just as they thought wolves stole sheep from the farms.

Where found • Lupin is widespread, numbering about two hundred species in Europe, Asia, and North and South America. Some varieties have poisonous properties, some are cultivated as ornamentals, and others are grown as fodder. Probably of eastern Mediterranean or Egyptian origin, some species are widely used today in Sicily and Italy as forage and as a soil supplement—lupin is ploughed in to enrich the soils. These varieties are also valued for their seeds.

Varieties of lupin are widespread in the Maritimes. The two species listed above are similar in appearance except that lupin is shorter than garden lupin. *L. nootkatensis* was introduced from the west, as its name suggests (Nootka Sound is in British Columbia), and has now become naturalized. It ranges from Alaska to Newfoundland and south into British Columbia in the west and New England in the east. It's also found in Europe and Japan. The range of *L. polyphyllus* is more restricted, but it too was introduced from the west.

Since the two species are so similar, they are combined in the information that follows.

Lupin seed pods

Habitat • roadsides, ditches, fields, and meadows

Lupin grow best in light shade or full sun.

Description • Lupins come in a beautiful range of colours—blue, mauve, pink, white, or yellow. The lupin grows on a narrow stalk that develops from the base upwards, reaching 40 cm to 1 m (15–30 in) in height. The flowers grow on cone-shaped spikes that are made up of scores of tiny flowers. The individual flowers are about 2 cm (about 1 in) in length, and bloom in June and July, opening over a period of several weeks on a single plant. The flowers last into the summer beyond most other spring plants. Each flower is a miniature version of the garden sweet pea, another member of the same family. The upper petal is erect, acting as a banner to potential pollinators and advertising the availability of nectar. Bees, especially bumblebees, are lupin's usual pollinators; they are attracted by lupin's subtle but sweet perfume. (Inside, as a cut flower, they look beautiful but some people find the scent unpleasant.)

Apart from the upper surface of the leaves, the entire plant is covered in soft hair. Each dark green leaf has six to nine leaflets that spread out flat like the fingers of a hand. The leaves are rounded or blunt at the tip, where they are widest. When the seed pods mature, they are covered in greyish hairs. You will find at least two seeds to a pod. Like other members of the Pea family, lupin's roots are studded with tiny nodules (see previous page).

Usable parts • not recommended for culinary or medicinal use; the plant does enrich the soil as it grows

Common uses • No culinary uses are recommended, but lupin is an important part of the landscape. The nodules, or tiny knobs, on lupin's large taproot contain a tiny living colony of bacteria. Nitrogen is needed in large amounts for healthy growth; it is one of the components in short supply in soils. The roots' bacteria convert nitrogen into a form that the roots can use in their search for food from the soil. The bacteria assist the lupin and benefit from the sugars—by-products of the photosynthesizing process—that the plant sends their way. As the plants die, they eventually release nitrogen into the soils for use by other plants. Thus they are a valuable enricher of the soils in which they grow.

Other uses • In 1917, at a conference of botanists in Europe, Dr. Albert Thomas hosted a banquet at which he celebrated and showed off the versatility of European lupins. The tablecloth and napkins were made from the fibre of lupin. Name tags, writing paper, envelopes—including the adhesive—and even gifts of soap were made from lupin. The entire meal was prepared from lupin—lupin steak roasted in lupin oil and seasoned with lupin extract, bread made with twenty percent lupin flour, lupin margarine, and lupin cheese (made from lupin albumen). The plant was toasted with its own liqueur, distilled from the plant. And coffee was served, made from the lupin beans.

Largely as a result of this event, varieties of lupin became a profitable crop in Russia and parts of Europe, but it is not yet cultivated to its potential. One product, lupin pasta, has been distributed commercially.

Caution • While varieties of lupin have been processed for use in food and beverages, the ingestion of any part of fresh lupin is believed to be toxic for humans and animals, though lupins appear to be rarely, if ever, consumed in sufficient amounts to be fatal. In addition, the species growing in this region may not contain enough toxins to be dangerous. Nevertheless, fresh lupin should be avoided until further research is carried out.

In gardens • I have been successful in transplanting lupins from roadsides by digging up the small plants in the spring; larger plants don't transplant well because of their large taproots. Once they become established, lupins will take care of themselves. It should be possible to gather seeds in the fall for spring planting, but the results may be unreliable as the seeds are very, very hard. They need to be filed, scraped, or soaked in water to facilitate their germination. You can certainly scatter the seeds in your garden in the fall and let the winter weather do its job naturally.

Lupins prefer well drained but not overly dry soils that are slightly acidic and rich in humus. You may enhance their colours by adding sulphates of ammonia and iron to the soil. Because lupins grow in nutrient-poor soils and light, sandy soils where few plants are able to thrive, they play an important role in ecological restoration. As a leguminous crop, they are able to obtain their own nitrogen from their bacterial partners and thus restore nitrogen to the soil for use by other plants. Lupins enrich the soil and make an excellent green manure.

Lupins are very hardy plants. Seeds from an arctic species (*L. arcticus*) unearthed in the Yukon from a mining operation were estimated at 10,000 years old. These were found frozen deep in the permafrost, but after forty-eight hours of soaking, some of them germinated! Garden cultivars are available now in nurseries. Look for varieties such as the Russell hybrids, the Minarette, and the dwarf variety, Gallery.

Marsh Marigold

Caltha palustris L.
Caltha natans Pall. • Crowfoot family

Other names • cowslip, waterdragon, meadow bouts, palsywort

Where found • Marsh marigold has a wide range, from Labrador across to Alaska and into the United States.

Habitat • slow, shallow water; wet marshes; swamps

Description • Appearing in May and June, the yellow flowers look somewhat like very large buttercups. The leaves are dark green and shiny, and are rounded to heart or kidney-shaped. The hollow stems grow from 30–60 cm (1–2 ft) tall.

Usable parts • flower buds, young leaves

Common uses • Marsh marigold must be carefully gathered and prepared. The young leaves can be used as a potherb or cooked green. Collect them before the flowers have gone, and remove the stalks. Boil them for thirty to forty-five minutes, changing the cooking water twice. Changing the water ensures that a bitter-tasting poisonous substance is removed from the leaves. The flower buds can be pickled as a substitute for capers. Soak them in salt water, then cook in two changes of water and pickle in hot vinegar. Do not drink the pickling juice.

Caution • Marsh marigold is poisonous if eaten raw. As explained above, boiling it expels the poisonous substance. Be careful when picking marsh marigolds not to gather parts of other plants; it grows in the same habitat as the poisonous water hemlock.

Mint
Mentha arvensis L. • Mint family

Other names • field mint, Canada mint

Where found • Wild mint can be elusive, but if you know where to look, the search will produce special rewards. Its name probably goes back to the Greek *minthe*. Worldwide there are many varieties of wild and cultivated mints.

Habitat • moist places—meadows, along ponds, lakeshores, riverbanks

Mint may be partly hidden by taller grasses. Sometimes you'll locate it first by its familiar scent or by spotting its tiny, bright flowers. The only times I came upon wild mint was when I wasn't looking for it at all. Both times the small plants were growing right at the edges of streams, partially hidden by other growth. They would have been visible from a canoe.

Description • Mint is easily recognized by its familiar aroma, which becomes strong when the leaves are crushed. Mint stems are square; the oblong jagged-edged leaves grow opposite each other. The plants grow up to about 45 cm (1 1/2 ft) high. The flowers appear in spring and summer. They are tiny (about 5 mm [1/4 in] across) and mauve in colour.

Usable parts • stem, leaves

Common uses • To dry mint plants, pick them when the flowers are open. Tie them in bunches

and hang them indoors or outdoors in an area that is well ventilated and free from direct sun. When the leaves are dry, crumble them and store in airtight jars. Mint can be used fresh, dried, or frozen in beverages and as a flavouring. A tea can be made from a spoonful of dried mint or a handful of chopped fresh mint. Steep the mint in a pot of boiling water for five minutes, then serve with lemon, a sweetener, or dried orange peel. Try adding some mint to your regular pot of tea. Mint teas are said to be good for digestion. Mint can also be used to flavour drinks, sauces, salad dressings, jellies, and fruit and vegetable salads, and is excellent served with new potatoes.

Other uses • Mint is sometimes added to regular smoking tobacco for flavour and aroma. In medieval times, it was used to stuff pillows or strewn on the floor to produce a scent when walked on. In England, it was grown in monastic gardens from the ninth century on.

Medicinal properties • In medieval times, mint was used to treat coughs and colds and was used as a restorative after a fainting or dizzy spell. It is antiseptic, antibacterial, and anti-inflammatory. Herbalists recommend it for headaches, digestive problems, flatulence, fevers, and colds.

In gardens • I keep a few of the plants growing at home in a bottle of water by carefully pulling out the plants with the roots attached. If you try gathering whole plants from a clump of mint, be sure to do it gently, and leave some plants behind.

Wild mint can be transplanted to a garden or planter. Choose an area that is similar to its habitat in the wild. Once established, however, mint is very hard to eliminate and can take over your garden with its very dense root system that other plants cannot penetrate. It's best to plant mint in a container, sunk into the ground to contain the roots. An old washtub works well.

Mullein

Verbascum L. species • Snapdragon family

Other names • Our Lady's flannel, candalaria, candlewort

Where found • Mullein is not native to the Americas (it is Eurasian in origin), although it is now widespread.

Habitat • Mullein prefers dry wasteland, gravelly banks, and disturbed soils, and is one of the first plants to colonize after a road or other development is built.

Description • A biennial, mullein overwinters as a flattened rosette that hugs the ground. In its second year, it develops into a tall, spear-like plant. The leaves are large, woolly, and covered with very fine hairs. These give a velvety feel but are also practical—they help to reduce evaporation and prevent insect damage. For hours sometimes, they will hold beads of dew or rainwater on their surface. As a leaf print, they make a splotchy impression because they are so soft and absorb a lot of ink (the same qualities that make it a great plant to have near an outhouse).

The leaves grow alternately on very thick and densely crowded spikes that can grow as high as 3–4 m (10–12 ft) tall, while the basal leaves lie flatly against the ground. The small, yellow flowers open all the way up the stalk in a random pattern all through the summer and provide easy access for pollinating insects. Mullein has a long taproot (like dandelion) to store energy and anchor the plant deep in the ground. It is a treat to watch a tall, mature, and sturdy mullein plant swaying in a strong wind.

Usable parts • leaves, flowers

Other uses • Mullein has a long and varied history of use. The Greeks made lamp wicks from the leaves, while the Romans dried the stalks, coated them with suet, and burned them as funeral torches, thus their name "candalaria."

The crumbled, dried leaves are a wonderful addition to a loose "smudge," or incense. The smoke provides a connection to the spirit world as well as cleansing and purifying our thoughts in this world. An early Irish herbal describes how carrying mullein preserved the wearer against enchantments and witchcraft. There is some history to this—Mercury gave it to Ulysses to defend him against the enchantment of Circe.

Medicinal properties • Fresh or dried, mullein leaves make an excellent poultice to stop the flow of blood in a wilderness emergency. It has been used externally to treat burns, wounds, skin diseases and ulcers. The leaves also contain a narcotic that can help to ease pain; patients who are treated with mullein may not need an anaesthetic at the hospital. Add a little water or vinegar to dried leaves to activate their healing properties.

Long used in folk medicine for pulmonary diseases as a tea, mullein leaves and flowers are used today as a tincture, tea, and oil for their mild sedative, antispasmodic, and other properties. In Ireland, it was developed as a treatment for tuberculosis by a Dr. Quinlan, based on a traditional Irish medicine. The leaves were boiled in milk (a handful of leaves to two pints of milk), then the liquid was strained and sweetened with honey. This was drunk at least once a day to soothe the lungs and build vitality.

As an herbal remedy, mullein has been used to treat asthma by inhaling the smoke from the burning leaves. It was also used to treat throat irritations, a deep, hollow cough, hay fever, emphysema, and bronchitis—the flowers contain a mucilage, or sticky sap, which is soothing. The prestigious German Commission E has approved mullein for treatment of respiratory catarrh, an inflammation of the mucous membranes.

Topically, it is applied as an oil for earache. You can buy mullein oil in small bottles in health food stores for this purpose. You can also make your own by soaking the flowers and the leaves in an extra-virgin olive oil (which doesn't go rancid). Then sieve the oil

to remove the plant remains. As an oil, mullein is also used as an emollient (a soothing agent) for wounds and itching, and can be rubbed into the chest area for respiratory problems. Gardeners are known to have laid a bruised, fresh leaf against their anus to give relief from piles and itching.

To make a tea that can be drunk three times a day, add about two tablespoons of flowers, or two or three leaves, to a cup of boiling water and let it steep about fifteen minutes.

Here is a formula to make a salve from mullein for use as a skin ointment for infection or bruising (but not for an open wound). Gather the flowers on a dry day so there is not much moisture in the leaves and let them dry on absorbent paper or over a piece of screening. Steep them in oil, then heat the oil and the leaves gently in a non-metal pot over a low temperature to extract their essence. Let this sit overnight so that the colour fades from the flowers. The next morning, heat it again, strain, and bottle the oil. If you want to make a solid medicine "cake," mix the oil with a little melted beeswax and pour it into a glass jar. A little less wax makes a paste; a little more will solidify the mixture. To use mullein as an infusion or decoction, carefully strain out the down or tiny hairs that cover the plant, as they are an irritant.

In gardens • Mullein, including its seeds, is extremely hardy. Mullein was grown in England's Mount Grace Priory, a fourteenth-century monastery. With the closing of the monastery by Henry VIII in 1539, herbs were no longer tended by the monks and ceased to grow. But in 1992, excavations of the priory's courtyard by archaeologists disturbed the earth. Several years later some of the seeds sprouted, including mullein (which was made into candlewicks at the time). That's a long gestation period—over five hundred years!

It's a plant I've seen just about everywhere I've travelled—in the Himalayas, Southeast Asia, Siberia, and across Canada. In fact, it grows wild throughout much of the world. In the archives of the British Museum, I was able to work with the earliest herbal in their collection, which has a beautifully detailed illustration of mullein dating back, if I remember correctly, to the 1400s. In my west coast garden on Saltspring Island, it was one of the first plants to establish itself after I'd put my road in, forming a tunnel of eight-foot-high plants. Its company there gave me great pleasure and continues to give me a sense of companionship in my travels wherever I go. Because of its ability to establish itself in a damaged or degraded area, I call it a restorative plant—it serves a very important role in repairing a landscape.

I also describe it as a plant that shows its whole life cycle on a single stem. The leaves at the bottom of the plant provide protection for insects and field mice whose droppings fertilize the soil. As the lower leaves desiccate and crumble into the ground to build a soil base, the stem grows taller and taller, producing flowers all the way to the tip and turning into a seed bank of thousands upon thousands of tiny black seeds. These provide food for many creatures and ensure that this very important plant spreads far and wide.

Nettle

Urtica L. species • Nettle family

Other names • stinging nettle, burning nettle, wood nettle, devil's plaything, Roman nettle.

The name nettle comes from the Old English Anglo-Saxon *noedl*, meaning "needle."

Where found • widespread throughout the world

Burning nettle (*Urtica urens* L.) is native to Europe. Stinging nettle (*Urtica dioica* L.) is classified as a noxious weed in North America.

Habitat • often found as a weed in waste places, along the sides of roads, or in other disturbed areas

Description • Nettles are known by the stinging hairs that cover their leaves and stems and exude an irritating acid when touched, a defense mechanism that protects the plant from herbivores. The genus name comes from the Latin *uro*, meaning "to burn." The species are not distinguished here as they are all easily recognizable and can be used in the same ways. Nettles are tall—2 m (6 ft) or more—spreading plants, and a beautiful darkish green in colour when they mature. The new leaves at the top of the plant are a lighter green. The leaves are oval or arrow-shaped, toothed, and taper to a point. By late June, nettle begins to flower and continues into September. Tiny flowers form clusters where the leaves meet the stem.

If you're wondering how nettles "work," the hairs are hollow and have fragile, sealed tips. At their other end, a tiny bulb filled with poison sits inside the epidermis of the leaf. When anything touches a leaf, the fragile tips are shattered and driven into the skin by the pressure of the touch. The same pressure pushes the base down, forcing the poison into the needle's tip, which is then injected into the skin. Cooking or drying the leaves neutralizes this effect.

Usable parts • leaves, stems

Common uses • Use the young leaves before the plant begins to flower. I extend the usable time frame of the leaves by snipping off the new growth that emerges at the top of the plant, delaying seeding. Some people go to great lengths to protect themselves when gathering nettles, wearing gloves and long sleeves. I gather them by going out with my cooking pot and a pair of pruning shears or even a pair of scissors. I just snip the leaves off and let them fall into my waiting pot.

Judy Goodsky, a Cree medicine woman, showed me how to gather nettles "bare-handed" and pop them into my mouth without getting stung—a great foraging technique if you like your food straight off

the stem. In short, if you touch nettle gently, you'll get stung. If you approach it with strong intention and handle it firmly, it will not harm you. Grab a leaf firmly between thumb and fingers, then fold it against itself into a little packet and place it on your tongue to soften it before chewing.

You can steam nettle leaves or boil them. If you boil them, don't throw away the "liquor," which is where the goodness lies. Sometimes I get my nettle to do double-duty: I use the liquor as a spring tonic and the leftover leaves in place of spinach in spanakopita or soup. The leaves can also be dried for later use. Today, they are commonly freeze-dried and put into capsules for the alternative health industry.

Old English and Welsh cookbooks gave recipes for nettle beer made from nettles and fresh ginger, or sometimes rhubarb, dandelion roots and leaves, and ground ginger. Nettles were also used in herb puddings, pies, pastries, dumplings, and porridge.

Other uses • As a fibre, which is made from the stems, woven nettle is extremely strong. As far back as the Bronze Age, nettle was used in textiles. It retains its strength through use but becomes softened, making it very comfortable against the skin. I have seen it made into duck nets in northern North America, a shaman's dress in Siberia, and scarves in the mountains of India. A biologist friend showed me a shirt made of nettle that he had purchased in Tibet, but I could not persuade him to part with it! To use the stems for fibre, gather them in the fall when they are fully dried, but before the rain or snow beats them down. That way you will have long, unbroken stalks. Pick or rub away the outer covering and twist the interior fibres into a twine. I make a two-strand twine by rolling the first strands on my thigh, then laying it over the second and rolling it. I keep them moist and aligned by dampening them in my mouth as I work along the strands.

Nettles were mixed with horse feed at the turn of the century to give the animals a beautiful, shiny coat. Commercially, nettle is a source of chlorophyll. The leaves produce a green dye, the roots a yellow dye. In medieval Europe, the seeds were considered an aphrodisiac. In one traditional method, nettle seeds were fed to hens to get them to lay eggs.

Medicinal properties • Nettle has a long history of medicinal use, going back in western culture to Hippocrates in 400 BC, who used the juice to treat bites and stings. The stems were struck against the limbs of paralyzed patients to stimulate the muscles. The Romans used it year-round as preventative medicine. Before contact, Native Americans were using it as a tea to help in pregnancy, childbirth, and nursing.

Homeopathically, nettle is used to treat asthma. Freeze-dried *Urtica dioica* has also been shown to relieve hay fever symptoms such as sneezing, sniffling, and runny nose and can be used effectively in place of antihistamines that may have considerable side effects, according to at least one controlled clinical trial. Nettle is one of the first green plants to show itself in the spring; thus, it is used commonly as an excellent, potassium-rich spring tonic.

In gardens • Nettle was cultivated as early as Anglo-Saxon times by monks in herb gardens and 'wort' beds. Do try to find a place for nettles in your garden, perhaps in a protected corner or at the end of a path, so that visitors will not brush up against them accidentally. I am very protective about my nettle patch and put a sign up saying "do not disturb" in case garden helpers think they are nuisance plants instead of the very valuable addition that they are. The traditional English garden left a wild space at the far end for the "fairies" or garden spirits. Nettles would be ideal plants for such a place.

Rudolph Steiner developed biodynamic sprays, including one made from nettle, to stimulate vegetative growth, especially during a dry spell. The same spray can also be used against aphids.

To make a compost activator, add nettles to the pile. Nettles rot fast and help break down other materials. They also have a high ammonia content, which speeds up fermentation. I also add them to a large plastic garbage can full of water. After they have broken down, I'm left with an excellent liquid fertilizer that goes directly into the soil.

Stinging nettle spreads both by rootstock and by seed. Gather the seeds when they are mature, at the end of the growing season.

Cautions • Some people may be allergic to nettle stings. The formic acid in the hairs may produce a reaction. While the occasional brush against a nettle plant produces a burning sensation and perhaps reddish swellings on the skin, I find neither lasts long. I try not to fall into nettle patches, however. If you have a tendency to do so, try to remember this children's rhyme: "Nettle out and dock in, Dock removes the nettle's sting." A classic remedy for a nettle burn or sting is to place a poultice of chewed plantain on the area, a very convenient remedy because the two plants are often found growing side by side. Rhubarb juice is another traditional treatment for nettle rash.

Partridgeberry

Mitchella repens L. • Madder family

Other names • twinberry, teaberry, squeakberry, one-berry, winter clover; called red berries in Labrador

The name partridgeberry is also sometimes applied to mountain cranberry (*Vaccinium vitis-idaea*), especially in Newfoundland and Labrador, where *Mitchella repens* is locally called two-eyed berry.

Where found • native to eastern North America from Newfoundland and into the U.S. as far south as Florida and Texas

Habitat • shady, mossy woodlands; hummocky pastures; damp banks

Description • Partridgeberry is a trailing plant that grows close to the ground. The tiny evergreen leaves are oval (nearly round) and shiny, growing opposite on the stem. They are leathery with short stalks and often have white veins. Four-petalled flowers grow in pairs at the end of the stem. Each is about 1 cm (1/2 in) long and whitish or purple/pink-tinged. They flower in June to July. The twin flowers form a single fruit—a bright red berry that sits on the end of the branch. Because they have a low fat content and don't rot easily, berries often stay on the stems throughout the winter and so are an important food source in the winter months for humans and wildlife.

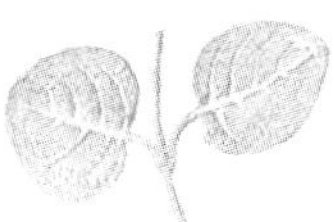

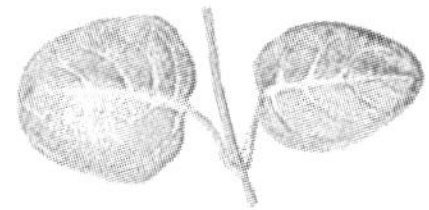

Common uses • The berry may be eaten fresh off the plant. Some people describe it as dry and tasteless, but it has always been important in Labrador and Newfoundland recipes for punch, fruit crisps, muffins, breads, puddings, jelly, salads, sauces, conserves, and jams. One traditional Newfoundland recipe mixes equal amounts of the pulp of partridgeberries with diced carrots to make a sweetened jam.

In gardens • Partridgeberry will form an excellent ground cover, especially in shady spots in your garden. It likes a rich "woodsy" soil (or a rock garden), like its native habitat. Like bearberry, partridgeberry roots from the joints and so is easily propagated in the same way, by pressing its stems into contact with the soil, or by dividing its roots.

Pineapple Weed

Matricaria matricarioides (Less.) Porter • Composite family

Other names • wild chamomile

Where found • Pineapple weed is an introduced plant that has become widespread in North America. It ranges from Alaska to Newfoundland and south into the United States.

Habitat • This common garden weed is plentiful along roadsides, in waste places and clearings, and around buildings. It grows in poor, gravelly soils. It will even push its way up through a crack in the sidewalk, cracking the cement as it grows. It is much tougher than it looks!

Description • Pineapple weed is fairly innocuous but easily recognizable. The yellow-green flower heads have no petals, and are about the size of small peas. They appear throughout the summer from June to September. Pineapple weed grows close to the ground, up to 20 cm (8 in) high. It may form large patches of ground cover, or the plants may be found singly. The leaves are finely dissected and look lacy and feathery. When crushed, the plant produces a very distinctive pineapple scent (although one friend described it as banana).

Caution • If you suffer from hay fever, you should probably avoid this plant.

Usable parts • flower heads, leaves, stems

Common uses • Pineapple weed is closely related to commercially grown chamomile. Chamomile tea has long been one of the most popular teas in the world. Traditionally it has been used for colds or as a sleep inducer and mild sedative to settle upset stomachs. Wild chamomile may be used similarly, and also makes a very pleasant-tasting tea. The flower heads can be used fresh or dried. Steep them in boiling water for five to ten minutes. The tea will be pale gold and pineapple scented. If you like, add a slice of orange or lemon. The leaves and stems are edible (and may be added to a tea), but some people find them unpleasant.

Other uses • Brew pineapple weed flowers, leaves, and stems as you would a tea, let it cool, and use as a hair rinse after shampooing. (This treatment is especially beneficial to blondes.) Add the stems and flowers to a hot bath for an aromatic soak by tying them in a net or cheesecloth bag and putting this straight into the water. Better still, make a tea out of them and add the liquid to the bathwater. You can also use dried pineapple weed to scent a drawer or linen cupboard, or to stuff a scented herb pillow.

In gardens • Pineapple weed is best left to its own devices. I like to leave it where I often find it—on footpaths, parking areas, and driveways. It can survive where other plants cannot. Stepping on it from time to time won't destroy the plant, and it will release a pleasant scent.

Pitcher Plant

Sarracenia purpurea L. • Pitcher Plant family

Other names • fiddler's trumpets, parrot beaks, purple pitcher plant

Where found • Pitcher plants—members of the *Sarracenia* genus—are only found in North America. Its nine species range in colour from burgundy to maroon to red (the leaves) and deep-red to yellow to maroon (the flowers). In the Maritimes, *S. purpurea* is especially common in Cape Breton. In 1954, it was made the floral emblem of Newfoundland.

Habitat • wet soils low in nutrients, swamps, damp alkaline, sandy areas

The pitcher plant thrives in the highly acidic conditions of bogs, the same preferred habitat as cranberry and other plants characteristic of the boreal bog. Bogs are peat-filled, moist, and unstable to walk on.

I came upon my first "wild" pitcher plant nestling deep amongst the Labrador tea, blueberries, sphagnum moss, and other ground cover of a blueberry bog on Eel River Bar reserve. Up to my waist in branches and sticks, I was taking pictures of the spiky skeletons of dead, deciduous trees that had been taken over by the bog and the black spruce that characteristically grows along a bog's perimeter. I looked down and into the centre of a three-foot tall pitcher plant! Thus began my fascination with this amazing plant.

Boreal bogs go back in time to when the glaciers retreated about 20,000 years ago. The glaciers scooped out depressions, which filled up with water. Lacking drainage, these eventually became bogs. Sediment washed in, built up, and the contents became highly acidic. Over time, a bog may fill in and become forest again so that no bog species remain, but this can take thousands of years. Pitcher plants have also evolved where wildfires were common.

Description • The pitcher plant is an insectivorous plant that captures and ingests protein-rich insects and spiders. As these are digested, the plant breaks them down into nitrogen-rich compounds that

are necessary for the plant's growth. The process ensures the plant a means of survival in nitrogen-poor soils.

A pitcher plant's leaves are large and funnel-shaped with wide lips. Look for the deep veins on the leaves. Look closely and you'll see tiny hairs pointing inwards to trap its meal and a sticky surface below to prevent escape. Its soup of digestive enzymes lies at the base of the leaf, where insects are eventually forced. Mature plants may have as many as twenty or more leaves, with some of them dying to make room for new growth.

Its single nodding flower head extends up into the air, perhaps a foot or more above the leaves, making it a difficult plant to photograph in its entirety. Five petals hang from the flower head. The variety I found has beautiful pigmentation—a soft blending of bright green and purply-red.

(Note: The tubular shape of the leaves meant a leaf print was impossible.)

Medicinal properties • Known as "Indian Cup Root" by First Nations, it was used by them to treat kidney ailments. The leaves were boiled and the resulting tea was used to wash sores and to treat children's rashes. Or, for the same ailments, a leaf was split and bound over the problem areas; this was thought to draw out "the bad stuff."

In gardens • The tall, stiff flower stems of the pitcher plant have become popular in floral arrangements, but they should not be gathered in the wild, since their habitat is becoming endangered as more and more bogs are filled in for agriculture and development. They are very easy to propagate, as several very large nurseries have found. Plants are now available commercially or can be grown from seed at home in peat pots. Collect fresh seed in the fall, after the capsules begin to turn brown. Top the peat pots with a layer of sphagnum moss and lightly press the seeds into it. Place the tray in the fridge for about sixty days, or leave it outside in cold weather. After you bring the tray inside or take it out of the fridge, seedlings will soon appear. Keep them moist all the time and plant them into larger containers when the first pitchers reach a half-inch in height.

To make a suitable habitat in your garden for transplants, choose an area that gets full sun or very light shade—pitcher plants need warm temperatures and high humidity through the summer. Scoop out a depression of at least a foot and a half in diameter and line it with a pond liner, either purchased from a garden supply place or made from a children's wading pool. Refill the depression with equal parts of coarse sand and peat moss, to about two inches from the rim so that excess water can drain. Keep it fairly moist by regular watering.

If you give the plants a needed dormant period of a few months each year (by removing them to a cool cellar or covered porch), they can be grown inside in a terrarium, kept moist by a covering of sphagnum moss. Mist them daily, keeping the pitchers two-thirds full of water. If you like, you can even "feed" your indoor plants with small insects or flies.

Plantain

Plantago major L. • Plantain family

Other names • soldier's herb, ratstail, broadleaf, door-yard plantain, greater plantain, round-leaved plantain, white man's foot, common plantain

Where found • There is uncertainty as to whether plantain is introduced or native to North America. Certainly it is a hardy and wide-ranging weed, and gardeners are aware of the firm grip it takes when it settles down in a lawn. There are several species of plantain in the region. All are similar in appearance to *Plantago major*, and all can be used in the same ways. It is among the easiest weeds to find and identify.

Habitat • common in gardens and lawns (where it is usually unwanted), along roadsides, in wastelands, and in fields

Description • Common plantain grows to a maximum height of about 25 cm (10 in). It has flat, elliptical, dark green, and dull-looking leaves that can grow up to 20 cm (8 in) long. These all radiate from a single base and are strongly ribbed lengthwise. Plantain's major rootstock is short and attaches itself firmly to the ground by many threadlike roots, making it difficult to uproot. Long, slender spikes of green-white flowers grow

in thick clusters at the end of the thick stems. The flowers appear all summer, from June through September. The plant is almost completely covered in fine hairs.

Usable parts • leaves

Common uses • Very young, tender leaves can be chopped and used raw in salads. As a potherb or cooked vegetable, plantain leaves can be used the way you would use spinach, and they make a good addition to soups and stews. The fibres running through the leaves, however, become tough and stringy as the leaves mature, and can cause problems. These fibres can be strained out by putting the cooked leaves through a sieve. Another idea is to remove older leaves from soups or stews before serving; this way you get the nourishment but not the nuisance. This is one plant that I find edible but not particularly palatable and so recommend it primarily as an emergency food source.

Other uses • An herbal tea can be made by soaking a spoonful of plantain leaves in a cup of boiling water to aid bladder problems and gastrointestinal ulcers. Herbals recommend crushing the leaves and applying them to cuts and scratches or using them for dressing wounds. The leaves may be applied whole to relieve the discomfort of insect stings, or the leaf can be chewed into a poultice and applied. Nettle stings can also be soothed in this way. If you're stung by nettles, look around for a plantain—they often grown close together. One herbal also recommends chewing on the rootstock to temporarily help toothache.

Wild Raspberry

Rubus L. species • Rose family

Other names and varieties • flowering raspberry, red raspberry, bramble, brambleberry, hindberry, dwarf raspberry, strawberry-raspberry.

There are a number of species of raspberry in the region; they are all edible and easily recognizable and so are not distinguished here. (The genus *Rubus* also includes cloudberry.)

Where found • Raspberry grows wild throughout Asia, Europe, and North America, where it is found from Alaska to Newfoundland and south into the United States. Even longtime Maritimers might be surprised at how many back alleys and garden fences are home to the wild raspberry. It remains common and widespread in the wild as well.

Habitat • clearings, roadsides, and open rock faces

Wild strawberry, fireweed, and raspberry are among the earliest pioneers or colonizers; these are "front line" in plant succession and ecological restoration after an area has been disturbed. In burned areas, they often form thickets. In the early days of North American settlement, wild raspberries were ploughed under to make agricultural land; it wasn't until the 1830s that they became important as a crop.

Description • Wild raspberry looks very much like the cultivated variety, but the fruits are smaller and less abundant on the canes. These slender, prickly stems can grow upright to about 2 m (7 ft) tall. The three-part leaves can be used at any time from spring to fall. The small white flowers appear from May to July, and the red fruit ripens in late July and can be picked till mid-August. The core of the berry pulls out easily when you pick raspberries ripe; blackberry cores, in comparison, stay with the fruit.

Usable parts • leaves, fruit, shoots, twigs

Common uses • To make a nice wild tea (a cloudy brew with a very pleasant, slightly sweet taste), pick the youngest, greenest leaves. Rinse them, then tear them up into small pieces, about a spoonful for each cup. Steep in boiling water, covered, for about five minutes.

The uses of fresh, dried, or frozen raspberries are well known. I usually eat them fresh-picked or with cream while I have the chance. While I never seem to have many left over for other purposes, they are excellent in pies, puddings, or other desserts such as raspberry fool, charlottes, and homemade sorbets or ice cream. Rich in pectin, they make excellent jams or jellies. I also make raspberry vinegar by adding a handful to a glass jar of fresh apple cider or white

balsamic vinegar for five or six days. Then I strain it and discard the fruit or serve it right away in a vinaigrette. In the spring, the young green shoots can be peeled and the inner stem eaten raw.

Raspberries are difficult to market commercially because they do not store or ship well fresh. Once picked, they need to be eaten right away or they can develop a mould. Look for them in local markets or U-pick farms, grow your own, or gather them in the wild. To freeze raspberries, lay them out in a shallow tray and put them in the freezer as soon after picking as you can—they are very fragile and crush easily and don't do well when they're placed more than one or two deep. Once they are frozen solid, you can tip them into freezer bags or containers.

Cautions • Raspberries are a favourite of black bears, so be careful when you're walking through the woods, and especially through a clear-cut. Make lots of noise by banging a couple of pans together so you don't surprise a foraging bear.

Medicinal properties • Herbals extol the virtues of various species of raspberry. A tea from the leaves is recommended as a gargle for sore throats and colds, and is a remedy for diarrhea. (The fresh fruits are a mild laxative and diuretic.)

The berries are rich in vitamin C, A, B1, and B2, and contain iron salts, calcium, phosphorous, and magnesium.

In gardens • Perennial rootstocks produce biennial stems, which do not branch or flower in their first year, but do in their second year. If you transplant some into your garden, pinch off the flowers in the first year they appear (i.e. their second year of growth) to strengthen the plants. About a dozen plants should produce enough berries for a family of four. Most commercial varieties today (red, black, purple, yellow, and even a French white variety) derive from a limited number of raspberry species, some native to North America and known for their hardiness.

Raspberries should be given a permanent sunny patch in your garden where they will be left undisturbed for ten years, perhaps as long as twenty. Off to the side or at the end of an established garden is best, though they should be planted away from trees that shade them and compete for water and nutrients. Make sure to keep raspberries away from any beds where a member of the potato family (which includes tomatoes, peppers, and eggplant) has been grown, as there is a possibility of soil-borne diseases to which raspberries are susceptible. Otherwise, raspberries are not fussy about their soil so will do well in most gardens. Their roots do go down about 1 m (3 ft) or more, so the soil needs to be deep. They need moisture but don't do well where their roots are soggy or waterlogged. An annual organic mulch of leaves, grass clippings, or sawdust works well to provide nutrients and keep down weeds but should be pulled away from the canes in the spring to allow the ground to warm.

Wild Rose

Rosa L. species • Rose family

Other names • bramble bush, briar, brier; various species are called swamp rose, rugose rose, sweetbriar, dog rose, and French rose

About fourteen species of wild rose grow in Canada. Many of these are common in the Maritimes. Cultivated varieties may be used similarly.

Where found • Wild rose shrubs are common throughout the region. They establish themselves quickly after a forest fire, transforming the burned area with colour and beauty, and providing life-support for returning insects, birds, and animals. Frequently taking over whole fields, they are a great joy to the eye and the nose, but their prickly and thorny stems can be a nuisance to the hiker.

Habitat • riverbanks, lakeshores, clearings, roadsides, often at the water's edge or at the edges of meadows

Description • The prickly shrubs grow up to 1 1/2 m (5 ft) high. Their pink, showy, fragrant flowers are common in the first summer months. After the flowers have bloomed and the petals have dropped, the fruits or hips are left. Green at first, these become fat and red by fall. They are packed with tiny, hairy seeds. (One of my ethnobotany students, Henry Smith, counted 129 seeds in a single hip.)

Usable parts • flower petals, hips, shoots, galls

Common uses • Roses have been cultivated for thousands of years as ornamentals and for their medicinal and culinary value. The oldest cultivated rose is thought to be a red rose, *Rosa gallica*. It may have been introduced by the Romans into Britain, and was also known to the Anglo-Saxons.

In the spring, the young, green shoots can be used as a potherb, or can be peeled and eaten raw. The petals, separated from their uncoloured and unpalatable base, can be candied, used in a tea (although a large quantity is needed), added to a salad, fruit punch, or even a sandwich as a garnish. Dried petals or the unopened buds are added to potpourri.

The hips can be used year-round, and thus provide an important emergency food source in the winter when they can be picked frozen. In fact, they are best picked after the first frost when they become their sweetest. They can be frozen, dried, or used fresh to make an excellent jelly, syrup, soup, or sweet, delicious tea, which has been packaged commercially for years. First Nations children call rosehips "itchy-bums," which an overdose of the seeds with their tiny, sliver-like hairs will cause! The hips can be steeped whole as a tea but for other uses, it's best to cut them open and scrape out the seeds. The seeds themselves are rich in vitamin E. After removing the hairs from the seeds by washing or scrubbing, the seeds can be used in cooking or as a vitamin.

To make tea, use fresh or dried hips, about a dozen per cup, and crush but do not split the hips open. You can also use the husks with the seeds and fibres removed. Steep the hips in boiling water—your tea will be a lovely, bright orange-red colour. Rosehip tea is readily available in tea-bag form; it's one of the main ingredients in many herbal teas.

Other uses • Rose galls are fuzzy growths that appear on the foliage of roses, especially the wild rose. I found a massive number of them one year, gathered them into a glass jar, and put several under my pillow to induce good dreams, the way medieval monks used them. Imagine my surprise when a few days later, the jar was active with minute wasps (*Cynipidae*, reddish or brown-coloured wasps in the gall wasp family). They lay their eggs in and around the flower buds in spring. When the eggs hatch, the tiny grubs colonize nearby leaf tissue, causing cell growth and a tangled mossy ball. The gall becomes food and protection for the tiny grubs, which eventually pupate in there, emerging in the spring as adults. At first, the galls are a beautiful mossy green; later they turn dark brown. This is when you can gather them to keep for your medicine cabinet, but make sure they hatch before you give one to someone to put under their pillow!

The petals produce an oil called attar of roses. It is used in cosmetics, soaps, fragrances, incense, and potpourri. Rosewater, available from Indian and Middle Eastern shops, is used as a skin softener and in salads and puddings.

Rose petal beads made from cooked, mashed, and moulded rose petals are a lovely way to use rose petals. You will need a lot, so you may need to supplement your wild rose petals with some garden varieties. The

use of rose petal beads originated in India, where they are used as mala beads and counted while saying prayers or mantras. By the third century, they were being made by eastern Christian monks and by the fifteenth century, the term "rosary" was coined. A string of rose petal beads will last for many years if you take care of it by keeping the beads dry—between uses store them in an airtight container. As you wear them around your wrist or neck, the warmth of your body will release their fragrance, and the oils from your skin will polish them over time. Wearing a strand has been said to be "like wearing angels around your neck." If you do want to restore the scent of old beads, rub some essential oil of rose into your palms and then rub the beads between your hands.

There are two basic ways to make your own rose petal beads. Either cook rose petals down in a stainless steel or enamel pot on top of your stove, or lay them out in a cool oven to dry over one or two days. You can dry, freeze, or boil the petals fresh (or use any combination of dried, frozen, or boiled petals). Gather all types of rose petals whenever they are in bloom; any colour will do, since they all turn black in the end. Make sure you have removed them from the stems and picked out all the debris. Place the petals in a pot with enough water to cover. Bring to a boil, then reduce to a simmer and cook for two to six hours. This can be done all at once or spread over two days. Stir occasionally. Put the mush into a blender in batches with enough water to purée it at the highest setting. Drain through cheesecloth to remove excess water. Now put the mash, a batch at a time, into an iron skillet—an old one is best as the iron will turn the mash black (and the mash will take the patina off a new cast iron pot). Use rubber gloves to prevent staining your hands. Cook the mash, covered to prevent a crust from forming, over the lowest heat you can over two or three days, turning it occasionally, until it is thick and creamy. Again, you can stagger the cooking time over several days. Or, you can cook it in the oven at 120°C (250°F) for a day or two.

On a tray, lay out a thick pad of paper towels on top of old towels or newspapers. Turn the mash onto the paper towels and cover it with another layer of paper towels. Let it cool and drain. Keep changing the under layer of towels or newspapers when the mash has soaked through. Knead in a few drops of rose essence to the mash—pure rose essence is expensive but it will last longer than the synthetic version.

You can freeze the mash in airtight containers or make the beads right away. For each bead, spoon out about half a teaspoon of mash. Wearing rubber gloves, mould it into a round bead shape. Cracks will show up later, so remove them by dipping the beads in a little water or rosewater (which you can buy in a bottle from Middle Eastern stores quite cheaply). Slide the beads onto brass wire (3/64 inch wire cut into 30 cm [12 in] lengths) by piercing the centres. Make sure the beads don't touch each other. Then suspend the wire over a box in a place the beads can dry at room temperature—too warm and they may go mouldy, too cold and they may freeze and crack. Turn the beads daily and slide them back and forth on the wire. Make sure they are completely dry before you remove them and store them in glass-covered jars with tightly fitting tops. Sand the wires with a fine steel wool before using them again.

Medicinal properties • Rosehips were distributed amongst the troops in World War Two for their high vitamin C content, as citrus fruit was unavailable. Rosehips are one of the best natural sources of vitamin C, containing ten to one hundred times more vitamin C than any other food, including oranges, and retaining it well even when dried and stored. They also contain calcium, iron, phosphorus, vitamin A, vitamin K, and the bioflavonoids. Today they can be bought whole, or in syrup, powder, or capsule form in health food stores.

Rose oil is known to include sedative, antispasmodic, local anaesthetic, antiseptic, antiparasitic, anti-inflammatory, laxative, and digestive properties. In aromatherapy, rose oil applied topically is used to treat nervousness, lift depression, relieve insomnia, and soothe anger and irritation. The hips, which contain vitamins C, A, B1, and B2, have been shown to work against gall and kidney stones; they are diuretic, astringent, and mildly laxative.

In gardens • Wild rose doesn't succumb to the problems of cultivated varieties, so it makes a welcome addition to my garden and I've found it easy to transplant. You can use the garden varieties for the same purposes as the wild.

Wild Sage

Artemisia L. species • Aster family (formerly the Composite family)

Other Names • prairie sagewort, pasture sage; wormwood (*A. absinthium* L.); mugwort or common wormwood (*A. vulgaris* L.)

Where found • More than one hundred species of wormwood grow worldwide. About fifty are found in North America. In the Maritimes, the most commonly used are *Artemisia absinthium* and *A. vulgaris*. Another member of the genus that gardeners will be familiar with is *A. stelleriana* Bess. (dusty miller or beach wormwood), which escaped from cultivation as an introduced plant from northeast Asia sometime since 1880 and is now naturalized.

Wild sage is strongly aromatic. Once you smell it, even from a distance, you will recognize its similarity to cultivated sage, although they are different plants. The aroma becomes even stronger if the leaves are crushed.

Habitat • cliffs, river terraces, dry open areas or slopes, disturbed soils

Description • Wild sage forms a lacy-looking mat of tiny but dense, silvery-grey-green leaves from a branching taproot. The leaves are very aromatic. The stems are slender and generally close to the ground, although these can grow 30–45 cm (1–1 1/2 ft) tall.

Usable parts • leaves

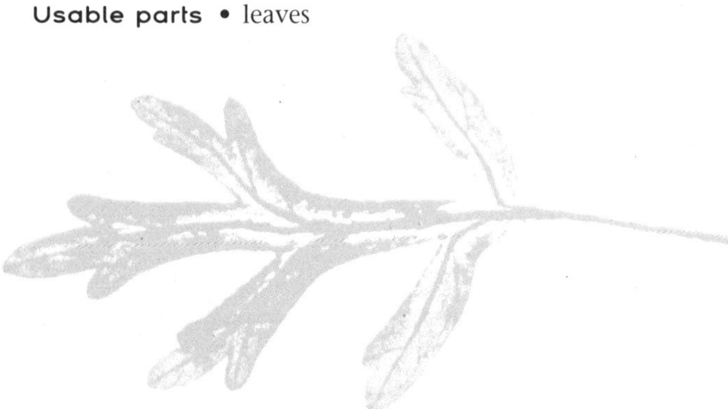

Common uses • To use as an herb, dried or fresh, carefully pick stalks off the plant (don't uproot it), tie them in a bundle, and hang them upside down in a dry, airy corner of your kitchen. Leave them there until you need them or crumble the dried leaves into glass jars to store. Use wild sage in stuffing for poultry, fish, or game, or as a seasoning for rice dishes.

Make an aromatic tea by steeping one or two spoonfuls of crushed, fresh, or dried leaves in boiling water. Add sugar and/or lemon juice to taste, or soak a slice of lemon or orange peel in the brewing water.

sea sage

Other uses • Toss some of the leaves into a campfire or place them on a hot stone and enjoy the fragrance. Wild sage can be added to sachets or herb pillows. The artemisias are used in Siberia and Mongolia and into North America as a sacred smudge, producing a scent said to be pleasing to the spirits. In North America, it can be bought as smudge sticks for use as an incense, or you can make your own by bundling the leaves together and wrapping them with cotton thread. Light your smudge stick from a candle or fire, then blow out the flame and let it smoke, resting it in an inflammable container—many people like to use an abalone shell.

Wild Sarsaparilla

Aralia nudicaulis L. • Ginseng family

Other names • American sarsaparilla, false sarsaparilla, rabbit's root, shotbush, small spikenard, wild licorice

Where found • Wild sarsaparilla ranges throughout the region's boreal forest from British Columbia to Newfoundland and south into the United States.

Habitat • moist or dry woods

Description • The plant looks stemless, since the actual stem is very short, and the leaf and flower stalks grow straight up from close to ground level; it reaches up to about 30 cm (12 in) in height. The leaf stalks divide into three parts, each usually bearing five oval leaves with serrated edges. The flowering stalks are also straight, shorter than the leaf stalks, and attach at the base of the plant. Each will usually produce one to several clusters of small, greenish-white flowers from June onwards and may carpet the forest floor in places. The fruits, which follow the flowers, are purplish to black in colour and each contains a single seed. Wild sarsaparilla has a long, aromatic root that lies flat at just below ground level.

Usable parts • roots

Common uses • Wild sarsaparilla is an ingredient in root beer. The roots are recommended as an aromatic nibble fresh out of the ground. They are also made into a tea, which is nice flavoured with mint or lemon. The fruit is edible, but not especially worthwhile.

Medicinal properties • Wild sarsaparilla and other plants in this family are best known for their properties as tonics and stimulants and have been used to treat chills, fevers, rheumatism, coughs, and to aid kidney and bladder ailments. The Mi'kmaq used the root to treat colds and influenza.

Shrubby Cinquefoil

Potentilla fruticosa L. • Rose family

The Latin name *potentilla* comes from *potens* because silverweed (*Potentilla anserina*), another species in the genus, was traditionally thought to have medicinal properties.

Where found • Shrubby cinquefoil is common in the region. It grows abundantly from the northern United States into the Arctic, extending into the tundra north of the treeline and into the alpine valleys of the Rocky Mountains.

Habitat • rocky places, muskeg, river banks, as well as in the most exposed and elevated places

Description • Shrubby cinquefoil is a dwarf, branching shrub that grows up to 1 1/2 m (about 5 ft) tall but is commonly shorter. Its reddish-brown bark is "shreddy" (rough in texture). Its many small leaves are made up of silky, toothless leaflets. The name cinquefoil, or "five-leaved," refers to the division of the leaves into five leaflets. In some places where the soil is rich, there may be up to seven.

Numerous yellow, five-petalled flowers appear from June through October, resembling small wild roses. One or several may grow near the tip of the branch. The flowers are followed by small, dry fruits. The flowers of shrubby cinquefoil are similar in appearance to silverweed and others of the many *Potentilla* species. One difference is that shrubby cinquefoil has toothless leaf segments whereas the leaves of silverweed are serrated. Also, the flowers and leaves of silverweed grow on separate stalks.

Usable parts • leaves

Common uses • Dry the leaves (you may also add the stems) or use them fresh to make a tea by pouring boiling water on them and letting them steep for five to ten minutes.

Medicinal properties • First Nations used the tea (made from the stem, leaves, and root) to treat a fever accompanied by a sore chest, sore arms and legs, and perhaps a cough.

In gardens • Shrubby cinquefoil is an excellent garden plant, as it is very hardy and virtually maintenance free. As it requires little watering, it does well in dry, hot weather, as well as in the cold. You may want to prune it occasionally to keep it from getting scraggly.

Silverweed

Potentilla anserina L. • Rose family

Other names • argentine, silver cinquefoil, goosegrass, silver feather, wild pansy

Silverweed is a member of the same genus as shrubby cinquefoil, but spreads with runners like the strawberry. It is difficult to distinguish the many species in this genus.

Where found • Silverweed prefers the cooler regions of Canada but extends south into the United States as far as New Mexico. Its east-west range extends across the continent from Alaska to Newfoundland.

Habitat • gravelly or sandy shores, riverbanks

Description • A low-trailing, easily noticeable plant whose flowers resemble tiny yellow roses on long stalks. The flowers are similar in appearance to shrubby cinquefoil, and appear in June through December, to be followed by small, dry-looking fruits. The feathery looking leaves have many-toothed leaflets. These are green above and silver-white below. The root fibres are fleshy and substantial.

Usable parts • roots, flowers

Common uses • In spring, the tubers taste a little like parsnips or sweet potatoes, and it's possible to eat them raw. They once provided an important winter food source for First Nations after the first heavy frosts. Dig them up before the ground freezes, then scrape and wash the roots. Roast, fry, or boil them for twenty minutes and serve them with butter. A traditional First Nations way of preparing them was to mash them with a stone before frying them in grease. They can be dried for storage and winter use, or you might still be able to dig them up in the winter on exposed banks.

Other uses • The flowers produce a bright yellow dye.

Medicinal properties • Silverweed tea is recommended in herbals to cure diarrhea and relieve cramps. Applied externally, it is considered helpful with skin problems and mouth sores.

Skunk Cabbage

Symplocarpus foetidus (L.) Salisb. • Arum family

Other names • swamp cabbage, eastern skunk cabbage

The Latin *foetidus* means "foul-smelling," referring to the distinctive stink of its leaves and flowers.

Where found • Skunk cabbage is a genus of only one species. It is native to eastern North America but no longer very common in the more populated areas. Also found in Japan and the eastern Asian mainland, skunk cabbage is an ancient plant. Like horsetail, it dates back in time to the very origins of flowering plants.

Habitat • boggy, swampy, mossy woods; wet areas in eastern deciduous forests

Skunk cabbage prefers deep, black, mucky soil and so is often found growing with sphagnum moss. Sometimes it can be seen in ditches at the side of the road. You'll find it in large patches and not as single plants since the large, marble-like seeds usually germinate close to the parent plant.

Description • Skunk cabbage is unmistakable in appearance and in scent. Most people describe the scent of the flowers and the leaves as putrid; however, I don't mind the smell. For me it is a welcome harbinger of spring. Skunk cabbage is one of the earliest plants to emerge from the forest floor, surfacing through the snow before other plants can get a start. It blooms while the buds of the trees in the forest canopy are still closed tightly so the plant captures the unfiltered energy of the sun. The yellow flower opens before the leaves and is cloaked by a yellowish, brownish, purplish leaf-like structure called a spathe. When the seeds form, the flower resembles an ear of corn.

Skunk cabbage flowers in May, after which the bright green leaves grow to 1 m (3 ft) or more in length—not many plants have leaves larger than skunk cabbage! Look for their very noticeable vein pattern. Amongst flowering plants, there are two major groups—dicots (most of the flowering plants), and monocots (which include skunk cabbage), and they can be distinguished most easily by their vein patterns.

Monocots
- Petals in multiples of three
- Have obvious, parallel leaf veins

Dicots
- Petals in multiples of four or five
- The net-like veins radiate out from a main central vein

The huge skunk cabbage roots go deep and are very hard to dislodge. If you pull on them, they'll pull back! It is better to put your hand into the muck and separate the segmented, worm-like roots with your fingers if you want to explore them. In colour, they are whitish, with a yellow tinge on the surface.

A close look at the leaves will show few, if any, indications of chewing by animals or insects. Skunk cabbage contains the chemical calcium oxalate, which produces a burning sensation on the lips. The odour is also a by-product of its system of defence against predators, and this odour attracts insects. Flies lay their eggs on the plant, which smells to them like rotting meat. As they do so, they accidentally pick up pollen. Flies and bees that feed on the tiny flowers also pollinate them.

A most unusual aspect of skunk cabbage is its ability to control its own tissue temperature. Thermogenic flowers produce their own heat (as humans do). Some, like skunk cabbage, can even regulate their temperature. An internal thermostat works even under the extreme conditions of below-freezing temperatures. As a result, skunk cabbage is able to maintain a temperature of up to 22°C (72°F), as much as 17°C (63°F) higher than the soil and the air temperatures that surround it. The plant uses oxygen and burns food at an enormous rate to keep its temperature so high. In fact, the male parts, on a gram-by-gram basis, produce more heat than any known species—plant or animal. Even the hummingbird can't match it!

The flowers generate enough heat to melt their way through ice and snow; thus they pollinate when the temperature of the air around them is just above freezing. One of the reasons the plant has developed this capacity is for survival. Another is to attract pollinating insects—the heat disperses the flowers' strong smell. Thus, skunk cabbage grows and matures faster than neighbouring wildflowers. Since the female's stigma mature a day earlier than the pollen-producing male stamens, it is in the flowers' best interests to keep insects around until the pollen is released.

Skunk cabbage has the ability to pull itself into the ground by contracting its roots, thus providing warmth for various small creatures that live or breed in them. Perhaps the heat is even a reward of sorts for pollinators to mate in a warm, protected environment.

Skunk cabbage plants can grow to a very old age, perhaps forever, or at least indefinitely, if they are not disturbed. Thus they may be older than the oldest trees that have survived in the region, perhaps even predating Columbus and other European explorers.

Common uses • The leaves make wonderful wrappings for forest gifts. For an outdoor feast, lay them on the floor of a pit oven for roasting potatoes and other root vegetables.

Skunk cabbage needs to be boiled in at least several changes of water before the leaves can be used as a cooked green and eaten. (Don't put any of the fresh plant into your mouth!) The leaves can also be dried and reconstituted for use in soups or stews, but be careful to dry them thoroughly and look for mould. You can dry the rootstocks, again very thoroughly, and pound them into a flour for thickening soups, or as an addition to other types of flours. Again, be careful not to put the fresh leaves or roots to your mouth directly or indirectly from your hands. The calcium oxalate will cause a burning sensation.

Caution • Be careful not to touch parts of the uncooked plant to your fingers or mouth. Use it only as a cooked vegetable. Thorough drying is the only other reliable way to remove the calcium oxalate.

Medicinal properties • The roots and the rhizomes, in a powder, liquid extract, or tincture, produce a resin and essential oil that has been used in the U.K. against dry cough, whooping cough, bronchial asthma, and the early stage of fevers. Both Maliseet and Mi'kmaq used it traditionally, the Mi'kmaq to treat headaches.

In gardens • Skunk cabbage won't find a suitable habitat in most gardens, but if you are lucky enough to have a patch in your forest, you can just expand the boundaries of your garden, perhaps even calling it a wild garden, as I do. You will find it a most interesting plant to visit from time to time, to warm up your fingers or as a reminder of just how recently humans inhabited the planet.

Spartina Grass

Spartina Schreber species • Grass family

Other names • cordgrass

The name probably comes from the Greek *spartine*, meaning "a cord."

Spartina patens (Ait.) Muhl.—salt hay or salt meadow cordgrass, prairie cordgrass

Spartina pectinata Link—broad leaf or freshwater cordgrass

Spartina alterniflora Loisel.—salt marsh cordgrass, smooth cordgrass

Where found • Of the sixteen cordgrasses in the genus (including some hybrids), only one is not from North America. *S. patens* and *S. alterniflora* are salt marsh plants (whereas *S. pectinata* grows in freshwater). In the region, the *Spartinas* range from British Columbia to Newfoundland and south into the United States. *S. alterniflora* has been recently introduced into Europe.

S. alterniflora has a bi-coastal profile now. On its native east coast, it is an important species and conserver of ecosystems. In the Pacific northwest, however, it has become an unwanted invasive species that is threatening its new habitat. It was introduced into the west coast of California a century ago, probably as a hitchhiker on the trains that transported gold prospectors and oysters between New York and New Jersey. The oysters were imported to replenish the dwindling west coast stocks affected by sediment runoff and pollution from gold mining in the Sierra Nevada. As the San Francisco Bay oyster industry moved north into Washington around 1940, spartina probably moved with it. Finally, in the 1970s, United States army engineers reintroduced it to San Francisco to try to reclaim the salt marshes. All of these factors led to the current situation on the west coast, where spartina is out of control.

Researchers at the University of California suggest that spartina hybridized with a native California cordgrass, producing taller, more aggressive plants that can self-pollinate (unlike the eastern spartina), thus encouraging their rapid spread. The hybrids grow closer to the low-water mark, thus producing so much biomass that it turns shallow waters into marshlands within a decade. Cordgrass is threatening to destroy oyster beds and bird sanctuaries and is resistant to commonly used methods of control such as herbicides. A new method being tried is that of biocontrol—introducing an aphid-like insect to kill the plant.

On the next page I discuss the east coast saltwater spartinas together.

Habitat • salt marshes and tidal flats along the coast

Description • Spartinas form large and often dense concentrations. The salt marsh is initiated by the arrival of a seed of spartina or the rafting in of a spartina plant dislodged from elsewhere. Spartina spreads quickly through its subterranean rhizome system. As it becomes denser, it traps fine sediments, including organic matter. As more is deposited, the sediment rises and eventually the habitat becomes terrestrial. Then, other less salt-tolerant grasses are able to take root and the marsh spreads.

Spartina's stiff stems stand erect, growing up to 1 1/2–2 m (5–6 ft) tall. The leaf blades grow to about 60 cm (2 ft) in length and about 2 cm (5/8 in) wide, with a folded tip, and are grey-green in colour. The flowers occur only on the undersides of the branches. Spartina blooms from early July through November.

Usable parts • roots, whole plant

Common uses • The *Spartinas* are essential to the salt marsh habitats. They prevent erosion of the salt marshes and help to define the high-water mark, which differentiates government-owned land from private. They are very efficient converters of inorganic materials into carbon-based foods for fish and wildlife. They can convert inorganic materials into organic more efficiently than crops such as wheat, and are among the contributors to the high productivity of the salt marsh. Snow geese forage on the roots that crabs and fish live among, feeding off the detritus spartina produces. As a feed for livestock, spartina is rich in nutrients; horses especially like it.

Spartina plants have the ability to secrete the excess salt that enters their root system or they would not otherwise be able to survive in such high saline conditions. They are very tough, as you can feel if you try to pull some away from their roots.

In gardens • Salt marsh ecosystems are extremely productive. Acre by acre, they can produce as much plant material as intensively farmed land, and the nutrients are replenished by the tides. These marshes provide fish and wildlife habitat (including sources of food for small fish and shellfish), filter out pollutants, and protect higher areas from flooding and soil erosion. Over some four hundred years, salt marshes between Nova Scotia and Massachusetts have been destroyed or degraded from farming, landfills, development, and dyking. We are beginning to understand the ecological value of salt marshes beyond their value to humans. An overall management plan is not yet in place but cooperation has begun between the United States and Canada to monitor salt marshes in the Gulf of Maine.

Salt marsh hay provides excellent winter mulch for the garden. It is free of weed seeds, and is airy, light, and doesn't mat down, so it holds air and warmth. Because it takes a long time to break down, you can use it over and over again.

Wild Strawberry

Fragaria L. species • Rose family

Where found • Widespread throughout the temperate parts of the northern hemisphere, from Alaska to Newfoundland and south into the United States; extends into South America. Some varieties are native to the Maritimes.

Fairly common, wild strawberry is one of the first plants to spring up through the blackened earth after a forest fire, along with raspberry and wild rose. You'll find it spreading its runners from the edge of the woods into the sunlight of a newly cut trail, or see it moving in soon after a clearcut, especially where the undergrowth and mosses have been torn up.

Habitat • open, sunny places, hillsides, and roadsides

Description • The wild strawberry plant looks similar to the cultivated varieties, except that the fruit and flowers are much smaller, and the fruit is much better tasting. The leaves are thin, light green, and divided into three coarsely toothed segments. Attractive five-petalled white flowers with yellow centres make the plant easy to identify in early summer. But strawberry plants grow so close to the ground that the berries, once they replace the flowers, may not be easily seen from above. If you get down and search under the leaves, you'll find plenty of berries, once they're in season—from mid-July through mid-August. New plants are produced by runners from the roots.

Usable parts • berries, leaves, runners

Common uses • The delicious wild strawberry is a treasure. The fruits are more flavourful than any cultivated variety though they can be used in the same way. To my mind, nothing beats a bowlful of ripe and fragrant wild strawberries. The berries are best eaten just after they've been picked but if you can hold back from consuming your berries on the spot, they can be frozen, dried, made into preserves or beverages, or used fresh in any number of desserts. To make a pleasant-tasting tea, add about a dozen fresh or dried leaves to a cup of boiling water. Steep for about five minutes and sieve.

Other uses • The juice of fresh berries rubbed on the teeth and left for a few minutes will remove discolouration. Rinse afterward with warm water that has a little baking soda added.

Medicinal properties • The medicinal qualities of the strawberry plant are widely known. (Check any drugstore for "Fowler's Extract of Wild Strawberry.") It has been used to treat wounds, diarrhea, dysentery, digestive upsets, and liver complaints. A gargle of strawberry tea is recommended in herbals for a sore mouth or throat. Strawberry juice is helpful to soothe mild sunburn. Strawberry leaf tea has been prescribed for bad nerves and anemia, and taken as a tonic after an illness. First Nations wind the runners around their fingers to dry them in a tidy coil, then store them for winter use in their medicine baskets.

In gardens • Gathered from the wild, strawberries are a summer treat. They like soils rich in organic matter and thrive in manure. Look for them in the grass in old pastures, old garden beds, and even

around tree stumps where the lawn has escaped the lawn mower. In the wild, they do best just left alone.

If you want to grow strawberries in your garden, you should know that they are heavy feeders and that they need regular additions of well-aged manure or compost. They also need potash to help them fruit. Mix wood ashes or seaweed meal into the soil, or pour on liquid seaweed, which will be most beneficial during strawberry's flowering period. In the garden, the plants cannot compete with weeds and grasses, which will strangle the plants if you do not take care.

The plants also need mulching. Traditionally, straw was used, but today farmers use black polyethylene which keeps the weeds down and also keeps in moisture. Holes are poked through the plastic sheeting and the strawberries are planted in the holes. I always build the soil up and plant in mounds for strawberries. They like the sun, and while they don't like being too dry, they don't like being wet either. Special strawberry barrels can make it possible to grow them vertically on a deck or patio.

Strawberries do need to be "renovated" each year to ensure good fruit production. Cut the new plants off at the runners and replant them in another bed so that you are always renewing your strawberries.

Sweet Gale

Myrica gale L. • Bayberry or Wax-myrtle family

Other names • bog myrtle, sweet willow, Dutch myrtle

Where found • A fairly common plant in the Maritimes and in the boreal forest, sweet gale is a Eurasian species. It ranges from Alaska to Labrador and south into the United States.

Habitat • along lakeshores; in wet, boggy areas; in shallow water or muskeg

Description • Sweet gale is a low, deciduous shrub, usually growing no taller than about 1 m (3 ft) tall. Its branches are bronze in colour. The leaves appear from late spring and last into fall. Measuring from 3–5 cm (1–2 in) long, they are highly scented, greyish-green in colour, and wedge-shaped, with slight toothing towards the tips. You'll find very fine hairs on both sides of the leaves. The half-inch-long cones are composed of small greenish-yellow nuts, which are sticky and aromatic. These can be gathered from late summer into the winter.

Usable parts • leaves, nuts

Common uses • Use the leaves to make a tea, or add them to soups and stews. The nuts can be used as a spice, the same way you would use sage; for example, sprinkle a few on a roast before it goes into the oven for a subtle flavour.

First Nations also used the buds in dyeing.

Sweetgrass

Hierochloë odorata (L.) Beauv. • Grass family

Other names • holy grass, vanilla grass, Seneca grass, Russian buffalo grass

Where found • Sweetgrass ranges from Alaska to Labrador and south into the United States, and is also found in Eurasia. It is used widely across North America for ceremonial uses by First Nations.

Habitat • Heavy, moist soils; upper areas of tidal marshes along the coast; wet meadows; slopes from moderate to subalpine elevations. It may also be found inland where it has been introduced. It grows well in bright sun to partial shade.

Description • Sweetgrass is a tall, erect perennial grass with a slight brown tinge. It spreads via creeping rhizomes that produce leafy stems up to about 60 cm (2 ft) tall, although most of the leaves are short. It flowers early in the season and may be gathered well into the fall. The flowers bloom into what may appear as a "pyramid cluster." It has a lovely vanilla fragrance, especially when crushed, and it is especially fragrant when dried.

Common uses • First Nations used it traditionally and still use it extensively wherever it is found.

Sweetgrass was used as a perfume for the body and sewn into clothing and other belongings for its long-lasting scent. The blades were soaked in water to make a hair wash and woven into baskets, such as the fancy baskets made by Mi'kmaq and Maliseet. It was also used in bedding, rugs, and mats. As an incense, it is still used to ward off negative influences. The longer leaves are braided into "sweetgrass braids," then allowed to dry and burned as a sacred smudge. Sometimes it was added, dried, to tobacco as a traditional offering. I was told by Larry Louie, a First Nations elder, that animals do not eat sweetgrass, so it is considered a very special plant.

Stephanie LaBillois of Eel River Bar showed me how to identify sweetgrass growing amongst other varieties of tall grasses in the ditches around her home. As you run your hands through the grasses, sweetgrass stands out because it catches the light differently—the light shines on one side of the blade. Then you run your hand down the long

blade and pull it up from the base to keep the stems together.

In France, sweetgrass leaves were boiled down to their essential oils, which were then used for flavouring beverages, tobacco, and candies.

Medicinal properties • Traditionally, it was used similarly to aspirin: chewed or held in the mouth, it numbed a toothache or soothed a sore throat; chewed into a poultice, it was applied to wounds as a painkiller and antiseptic. The smoke was inhaled as a medicine. The root was boiled to make a tea for treating colds and coughs. The soaked stems were applied topically to soothe wind-chapped skin and sunburn. For horses, sweetgrass was used in the healing of saddle sores. It was also fed to them to increase their endurance.

The bitter taste of sweetgrass comes from coumarin, a plant chemical that appears in a number of vanilla-smelling plants. Coumarin is used in orthodox medicine as a powerful anticoagulant and so is used to prevent blood clotting. Thus, sweetgrass should not be ingested during pregnancy or breastfeeding.

In gardens • The sweetgrass seed is hard to find, but many nurseries carry plants, including some mail-order companies in Canada and the United States. Sweetgrass rarely grows in pure stands; it is usually found among other grasses, so you may want to devote a section of your garden to grasses. If you do find a healthy stand with a lot of sweetgrass, you can propagate it through division of the rhizomes (since few will germinate from seed). Plant it in soils that simulate its habitat in the wild—cool, moist soils with lots of organic matter. Since sweetgrass is endangered in some parts of the Northeast as a result of habitat degradation and competition from introduced species, you will be helping to ensure its viability.

Tansy

Tanacetum vulgare L. • Aster family (formerly the Composite family)

Other names • golden buttons, garden tansy
The Latin name *vulgare* means "common."

Where found • This is a widespread genus. *T. vulgare* ranges from British Columbia to Newfoundland. It is native to Britain and was known to the Anglo-Saxons. The Latin *tanacetum* derives from the Greek word for immortality, referring to the long-lasting nature of the flowers when dried; it also makes reference to the ancient use of tansy in embalming and preserving corpses.

Habitat • roadsides, orchards, fields, along salt marshes and roads

Description • Tansy flowers in July and August, producing clusters of yellow, button-like flowers. These droop slightly from the stems that grow up to 60–90 cm (2–3 ft) tall. The plants are found growing in clumps from stout rootstocks. The lacy-looking leaves are finely divided, aromatic, and bitter-tasting. The plant gives off a strong odour.

Usable parts • leaves

Common uses • A few leaves were once added to sauces, salads, and omelettes in Britain. Another traditional use was in pancakes to mark the end of the fast undertaken at Lent. They were also added to custards, creams, and possets (a type of spiced drink), and other dishes as a substitute for cinnamon and nutmeg. Stew the leaves with rhubarb for an unusual taste and sweeten with stevia, a natural sweetener that has none of the negative properties of sugar.

Other uses • In the Middle Ages, tansy was a popular strewing herb, scattered on the floors of homes as a disinfectant and insect repellent. Before refrigeration, the leaves were also used as a wrapping for meats to preserve them and keep flies away. Tansy makes a pretty addition to a wildflower bouquet, fresh or dried. It is also added dried to potpourri. The flowers produce an orangey-yellow dye.

Caution • The herb should not be taken during pregnancy and should not be taken in excessive amounts.

Medicinal properties • Tansy has a long history of medicinal properties. Externally, it was applied to sprains, swellings, bruises, and varicose veins. Taken internally, it was given to aid digestion, expel worms, promote menstruation, relieve flatulence, and treat gout. As a tea, it was prescribed as a general tonic. It was also used to promote the appetite, treat jaundice, strengthen the heart, and reduce blood pressure.

In gardens • Tansy will often find its way into your garden amongst other wildflowers and will come up year after year. It makes an easy-care addition to a cutting garden or to a patch you plan to dry, perhaps scattered amongst straw flowers, which also dry easily. To dry tansy, hang bunches upside down in a well-ventilated area out of the sun.

Tobacco

Nicotiana rustica L. • Solanaceae or Nightshade family
Nicotiana tabacum L.

Other names • *Nicotiana rustica*—known as Aztec tobacco, picietl, mapacho, Zuni tobacco, and Native tobacco—is one variety of about twenty-two species of *Nicotiana*.

Where found • Archaeologists have documented the use of *N. rustica* in pre-contact South America. It is an extremely potent variety of tobacco still cultivated in the Amazon and used by South American shamans as an entheogen (a psychoactive substance) to bring about a spiritual experience. It contains nicotine in high concentrations, up to twenty times higher than other North American varieties such as *N. tabacum*. *N. rustica* is probably the variety that First Nations, especially the Mi'kmaq, were cultivating before contact with Europeans in the Northeast. Today *N. rustica* is under cultivation in India and Turkey.

The tobacco genus is related to other genera, including two that have been domesticated to provide tomatoes and potatoes. The family also includes the highly toxic and even poisonous nightshades, henbanes, and daturas.

Habitat • *N. rustica* is native to the southwest United States, Mexico, and parts of South America. It can also be grown throughout the United States and parts of Canada.

Description • *N. rustica* is an annual that grows to about 1 1/2 m (5 ft) tall. The leaves alternate on the stem, growing larger at the base and reducing gradually towards the top of the stem. Its pale yellow flowers are trumpet-shaped, approximately 2.5 cm (1 in) in length. They exude a somewhat unpleasant odour. The tiny, numerous seeds are dark in colour and form in capsules.

Tobacco was introduced into western Europe from the Americas. Early interest in tobacco amongst Europeans was first as a pretty "adornment" to gardens, and then almost entirely for its medicinal qualities. As early as the sixteenth century, plants from the Americas began to be included in herbals, which served as health manuals. These recommended its use for asthma, cancerous sores, the plague, and labor pains, in the form of enemas and poultices as well as for the smoke. By 1604, a debate was raging about its use, and the English king, James I, condemned its effects on health and social behaviour; by 1624, he had reversed his initial decision to support colonial tobacco growers. Tobacco is the traditional gift in indigenous communities. The leaves are dried and then wrapped in a cloth to be given as a traditional sacred offering.

Medicinal properties • Today, science is once again recognizing nicotine's "virtues" without recommending that people take up smoking. Research is still in its preliminary stages but the Duke University Medical Center is finding that nicotine produces beneficial effects on the brain by enhancing cognitive functioning and assisting with Parkinson's, schizophrenia, and Alzheimer's. Whether such promise ever sees mainstream use is unlikely, however, because nicotine as a natural substance cannot be patented and thus pharmaceutical companies are not interested in investing in clinical trials.

Tobacco in First Nations' ceremonies was a sacred medicine. It was offered as a gift to the Creator, to give thanks to Mother Earth, to seek guidance and protection, and to ask for discipline and spiritual strength. It kept the sacred balance between the spirit world and the natural world. Like all medicines, it could be both beneficial (if used with respect and the right intent) and harmful (if used in the wrong way).

In ceremonies tobacco may be burned as an offering or smoked in a pipe. The prayers of the elder, as well as the prayers of the giver, are in the smoke that rises from the fire and travels to the spirit world, linking the two worlds. Smoke, I was told, is one of the things that can permeate the very thin membrane between the worlds. Before participation in a sweat lodge, participants abstain from tobacco, alcohol, and caffeine, although tobacco is burned in the sweat lodge ceremonies.

Tobacco is not always burned, but may be offered in dried form before or after the killing of a deer or other animal. It may be given to elders at the beginning of a ceremony or as a sign of respect. Or it may be placed or sprinkled on the ground before collecting plants for medicine, again as a sign of respect and as a gift to the Creator. The Mi'kmaq say that the instructions for using tobacco were given to them through Glooscap, the messenger of the Great Spirit Creator. Today, First Nations following the traditional way will not buy or sell native tobacco. I was given a plant by Ray Solomon on the understanding that I would share it with others as he shared it with me.

As a medicine, First Nations used it to treat earaches, fever, sore eyes, insect bites, burns, asthma, and problems of the bowel. In South America it was used as an enema. Amongst First Nations, "tobacco" also refers to a mixture of plants such as bearberry leaves, sweetgrass, Labrador tea, red willow, and wild sage that was used in the same way as tobacco and sometimes mixed with it.

In gardens • Native tobacco can be easily grown in gardens or in containers inside the house. While I don't encourage smoking, homegrown tobacco is missing the chemical additives that go into cigarettes. You may choose to sprinkle a little in your garden before you harvest your crops as an offering to the spirits or to nature, as I was taught to do by First Nations.

The high concentration of nicotine in *N. rustica* makes it a useful organic pesticide.

Turtlehead

Chelone glabra L. • Snapdragon or Figwort family

Other names • balmony, cod mouth, shell flower, snakehead, white turtlehead, bitterherb

The common name comes from the resemblance of the corolla, or flower petals, to the head of a turtle or reptile. The Greek *chelone* means "tortoise." In Greek mythology, the nymph Chelone refused to attend the wedding of Hera and Zeus. In punishment for her insolence, the gods changed her into a turtle.

Where found • A native perennial, turtlehead's range is restricted to eastern North America, from Newfoundland to Minnesota and south into Georgia and Alabama.

Habitat • wet woods and meadows, marshy areas, along streams, flood plains, and wetlands.

The roots are often sitting in water. Turtlehead is actually an indicator species of fens, which have slightly acidic to alkaline soils.

Description • Turtlehead is named for its almost scentless, tube-like flowers, each about 3 cm (1 1/4 in) long, with a mouth at the outside end resembling that of a turtle or snake. The flowers are white and may have a pinkish, reddish, or purpley blush. They are closely packed on a spike (that is four-angled and occasionally branched) from 60–90 cm (2–3 ft) tall and may grow singly or in clumps. About mid-June, they begin blooming from the bottom of the spike and continuing to the top. I found them still blooming but just beginning to brown at the edges in the Sackville Waterfowl Park at the beginning of September.

An ovoid seed capsule containing several flattened, broadly winged seeds replaces each flower. The leaves appear to be stalkless, attaching to the stem opposite one another. Each pair of leaves rotates ninety degrees from the pair below it. They are hairless and lance-shaped with serrated edges. The lower leaves may be broadly oval and narrowing towards the base. Turtlehead's root system is a taproot and rhizomes.

Usable parts • whole plant but especially the leaves (should be gathered during the flowering period)

Common uses • Because it is a true native species, turtlehead and its habitat should be preserved for our region's biodiversity, apart from its value to us as a medicine.

Caution • Pregnant women should not use turtlehead under any circumstances. In the Mi'kmaq/Maliseet pharmacopoeia, it was taken to abort and said to be efficient up to four or five months into pregnancy.

Medicinal properties • The extreme bitterness of turtlehead's leaves made it very useful in the 1800s for treating such diseases as jaundice, fever, hepatitis, and eruptions of the skin, among other complaints. One New England writer described it as having long been known as a laxative and tonic and noted that it was used to treat children with worms. First Nations were known to use it as a decoction of the whole plant to treat hemorrhoids, sores, and other kinds of skin "eruptions." Today, homeopathic products use turtlehead in combination with other plants to treat nausea, flatulence, and indigestion.

In gardens • If you have a wet area or pond in your garden, you should be able to cultivate turtlehead plants (space the plants about a foot apart) but make sure you water the plants during dry spells as they need wet to moist conditions in partial to full sun. The soil should have some organic matter and be slightly sandy and acidic. Like other wetland species, turtlehead's natural habitat is increasingly being taken over by development. You will be contributing to the survival of a native species that was growing here and was in use by First Nations before Europeans arrived. (You'll also be contributing to the survival of its companion butterfly.) Buy the plants from a reputable nursery that does not gather from the wild, though.

Yellow Pond Lily

Nuphar variegata Durand / *Nuphar X rubrodisca* Morong
Nuphar microphylla (Pers.) Fern.

White Water Lily

Nymphaea odorata Aiton• Water Lily family

Where found • Yellow pond lilies, as well as the white water lily, grow in the Northeast. They are combined here because they are similar in habitats and uses.

Yellow pond lilies:
Nuphar variegata ranges from the Yukon to Labrador and into the United States. It is also found on Sable Island.

Nuphar microphylla is more restricted, from Manitoba to the Maritimes and into the United States.

Nuphar X rubrodisca is even more restricted in its range, found only between Minnesota and Nova Scotia.

White water lily:
Nymphaea odorata Aiton
Very common in the region. Ranges from Manitoba to Newfoundland and south into the United States as far as Florida.

Habitat • ponds, pools, gentle streams

Description • These aquatic plants have long, floating flower stems and leaf stalks; their true stems are underground. The yellow pond lily has a thick, cylindrical root and a single, yellow, globe-like flower. The white water lily is smaller and more delicate than yellow pond lily and grows in shallower water. Its white unscented flowers grow to about 5 cm (2 in) across. Roots can be gathered any time of the year; the seeds, in late summer and fall.

Usable parts
Yellow pond lily—roots, seeds
White water lily—young leaves, flower buds

Common uses • While always remembering that the aquatic lilies may become too rare to be a regular food resource, you can derive good nourishment from them if the need does arise. The roots of the yellow pond lily are recommended as a starchy vegetable. They are peeled, then boiled or roasted. If you dislike the flavour, try boiling them in several changes of water. Or you may be one of those people who find them impossible to eat. In this case, "edibility" becomes a question of taste or survival. First Nations in the region used the roots to make flour.

The seeds of the yellow pond lily can be taken out of their pods and boiled, added to soups, or roasted and then ground into a cereal or bread additive. They can also be used as popping corn and are rich in oil, protein, and starch. (When the seeds are older, their shells harden and must be soaked in hot water or roasted before the centres can be removed.) The leaves of yellow pond lilies can be used in place of aluminum foil as a wrapping for roasting or baking vegetables.

The unfurling leaves of the white water lily can be used in emergencies as a cooked green, as can the closed flower buds. Boil them for five to ten minutes.

Medicinal properties • The roots or rhizomes were used as a medicine to reduce swellings by the Mi'kmaq and Maliseet.

Poisonous Plants

There is no truth to the common idea that nothing poisonous grows in the Northeast. While no true arctic plant, berry, or root is considered poisonous, the story is very different south of the treeline, where most of the Northeast lies. As you become familiar with the edible wild plants, you should also learn what to avoid. The following pages list the poisonous plants you'll most often encounter on your rambles, and explain how to identify them. As well, many plants in this book, while not listed as poisonous, are accompanied by cautioning notes. Readers should take these seriously to avoid needless difficulties. In general, poisonous plants are of two kinds: those that cause illness if ingested and those that are irritants to the skin. Both kinds are found in the Northeast.

The most widely known plants that irritate the skin—poison hemlock (*Conium maculatum*), poison ivy (*Toxicodendron* species), and poison sumac (*Toxicodendron vernix*)—do grow in the region, as do other poisonous plants with irritating properties such as the common buttercup (*Ranunculus* species). Buttercup "juice" can cause skin blisters if touched and intestinal inflammation if taken internally.

Of the edible species in this book, cow parsnip (*Heracleum lanatum*) is one that can cause skin irritation and should be handled with care. In fact, it should be avoided entirely unless you are absolutely certain of its identification since it can easily be confused with a poisonous species, water hemlock (*Cicuta maculata*). It should be remembered that some people are more susceptible than others to irritants, and that some will not be affected at all.

Many plants commonly used to make food and beverages are dangerous in excessive amounts, though harmless if used in everyday or limited quantities. For example, Labrador tea taken in large doses may cause headaches, cramps, or other complaints in some people. Some edible plants contain chemicals closely allied to common medicinal compounds. Willow, for example, contains a substance whose synthetic compound is aspirin. Such plants, and others, are accompanied by cautioning notes in the pages of the field guide. To be safe, try such plants in small quantities at first, until you are sure of your own reaction to them.

Some plants must be processed, usually by cooking, to render them edible. Lichens, for example, should be boiled and treated with baking soda to remove their toxic substances. Marsh marigold in its raw state is poisonous, though the leaves look as if they would make a delicious salad. (Cooked, in fact, the leaves are nourishing and widely used as a green.) Other plants are edible only in part, while the remainder of the plant should be avoided. The bristly centres of rosehips, for example, are best removed before the hips are used.

Certain plants are extremely poisonous and can cause death. They are particularly dangerous if they can be mistaken for edible species discussed in this book. You should learn to recognize the following plants and avoid them for culinary use: water hemlock (*Cicuta maculata*); baneberry species, red and white (*Actaea* species); and wild sweet pea (*Lathyrus maritimus*).

Where possible, this field guide points out toxic properties, but readers should keep in mind that relatively little is known about the toxicity of wild plants. New information becomes available all the time; don't rely on non-current sources, and be prepared for conflicting information. This book does not list all the known (or potentially) toxic plants that you may encounter in the Northeast. For more comprehensive coverage, readers are referred to John Kingsbury's books on poisonous plants.

Another note of caution • Plants intended for food and medicinal use should not be gathered from any area that has been sprayed to control weeds or insects, or where there is contaminated water. Be careful even if you are gathering for basketry or dyeing. In fact, the plants you gather for any purpose may need to be washed in clean water before use, particularly if they come from a roadside or inhabited area.

Water Hemlock

Cicuta L. species
Parsley family

Other names • poison parsnip, beaver poison, muskrat weed, musquash root, poison hemlock, wild carrot, false parsley

Plants of the genus *Cicuta* are considered to be the most violently poisonous in North America. They can cause convulsions, diarrhea, stomach cramps, delirium, and death.

How to recognize • Water hemlock can be particularly difficult to identify, partly because its leaves take different forms depending on habitat. A further difficulty is the fact that the plant shows variation at different stages of its growth, and is especially dangerous since it may grow intermingled with edible plants that it resembles, such as cow parsnip (*Heracleum* species) or other members of the Parsley family. For this reason, a warning appears in the field guide description of cow parsnip, and its use as a food plant is discouraged unless its identity is absolutely certain.

Habitat • All species of *Cicuta* prefer similar habitats, which must be moist or wet for at least part of the year: they are found in moist meadows, shallow water, marshy areas, and swamps.

Description • Water hemlock plants usually grow 1–2 m (3–6 ft) high, although they can grow up to 3 m (10 ft). The white, umbrella-shaped, flat-topped flower clusters measure about 5–12 cm (2–5 in) across. They appear from June through August. The leaves alternate on the stem: the lower ones branch, and are often divided into three, their leaflets having toothed edges. The stem can be quite stout. It thickens where it joins the roots, and it is hard to differentiate from the roots. The roots are usually fleshy, tuberous, and branching. They have an odour of cut parsnips or carrots and are often, dangerously so, mistaken for wild parsnips.

Caution • The roots in particular (but also the other parts) are very poisonous. A very small amount can be fatal.

Wild Peas & Vetches

Lathyrus L. species & *Vicia* L. species
Pea family

Other names • wild pea, licorice root, wild licorice

Lathyrus species include beach pea, wild pea, meadow pea, and everlasting pea; *Vicia* species include the vetches, such as hedge vetch, cultivated vetch, hairy vetch, and slender vetch.

How to recognize • Wild peas and vetches are widespread and include almost three hundred species worldwide between them. They are similar in leaf arrangement. The vetches have white or blue flowers; the wild peas have red-purple to white or yellow flowers.

Some members of the Pea or Bean family resemble licorice root (*Hedysarum alpinum*). Foragers are advised to be cautious when gathering licorice root to avoid collecting a wild pea by mistake. Many published sources confuse the two; *Hedysarum mackenzii*, a northern variety, is often referred to as "wild licorice" or "licorice root." Some sources classify *H. Mackenzii* as edible: the roots have a sweet licorice-like taste. Others indicate that the root is slightly toxic. It's possible that references to *H. Mackenzii* as poisonous derive from a single instance cited in 1852. It's also possible that its toxicity varies in different locations. In view of this, *H. Mackenzii* is not recommended as a food plant. While *Lathyrus* and *Vicia* species make pretty ornamentals in a wild flower bouquet, these should not be eaten or confused with an edible species.

Habitat • Lakeshores, riverbanks, gravelly soils; often a troublesome weed in gardens and cultivated fields

Description • Wild pea plants and vetches are climbing or sprawling vine-like plants; some have tendrils by which they spread along fences or other plants. The seed pods that follow the flowers are longish and contain two or more seeds depending on the species. They resemble cultivated pea pods until they ripen and turn brown and dry on the stems.

Red Baneberry

Actaea rubra (Ait.) Willd.
Crowfoot family

Other names • doll's eyes

How to recognize • Baneberry is an attractive plant whose berry-like fruits are highly poisonous. They are capable of causing severe gastroenteritis; European species have apparently caused deaths among children.

Both white and red baneberry grow in the region and in other parts of North America. Red baneberry is wide-ranging in the boreal forest.

Habitat • moist, subalpine meadows

Description • A plant may have one or several stems, from 45–100 cm (18–40 in) in height. Its leaves are thin but generous in size with sharply toothed segments. The white flowers are tiny and numerous. The root is a thick, fleshy rhizome. The poisonous berry-like fruits are bright red, although occasionally pink berries are found.

Other Plants to Avoid

Some very common Northeastern plants are best left alone or used cautiously.

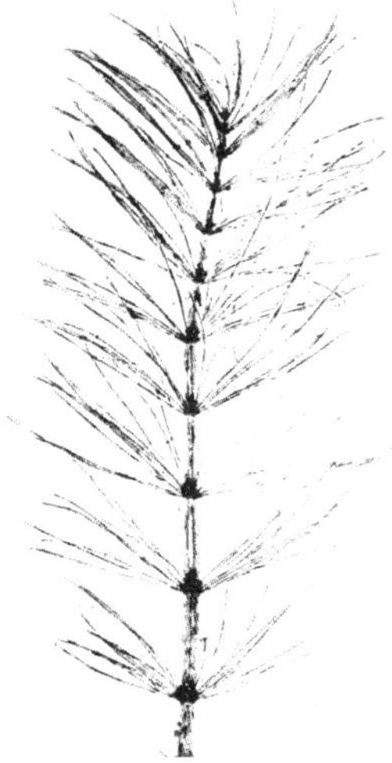

Horsetails (*Equisetum* species)
Horsetails have been reported as being responsible for the death of horses, cattle, and sheep. Studies of their effects on people are not conclusive, but raw horsetails in quantity could be fatal to humans. In reasonable quantities, however, the shoots are considered edible, as this book discusses.

Ferns
Many books on edible wild plants recommend ferns. While some species have been identified as edible, many are now being reconsidered as potentially dangerous. The ostrich fern—*Matteucia struthiopteris* (L.) Todaro—is considered to be non-poisonous and is covered in this book under "fiddlehead." Other ferns growing in the Northeast should be avoided for food purposes, at least until more is known about their properties.

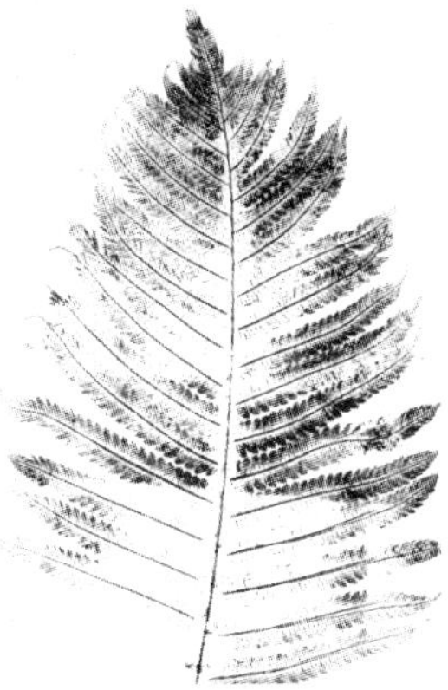

Mushrooms
Because some of the Northeastern species are poisonous and identification is often difficult, it's recommended that you be particularly cautious in using mushrooms, even if you use reliable sources. They are beyond the scope of this book.

Part Three: Recipes

A note on CLEANING, DRYING, and STORING wild plants

Your wild harvest should be carefully picked over for dead or damaged leaves and berries, bits of stems, and clinging insects. If necessary, wash plants thoroughly in cold water before cooking or drying. Berries are better left unwashed, if possible.

To dry plant parts for future use, pat them in a towel after rinsing, and spread them on newspapers, keeping the layer thin and well spaced to allow free air circulation. Split thick stems; separate multiple flower heads or berries that grow in bunches.

Lay the newspaper flat in a dry, warm, dust-free place out of direct sunlight and wait until the plants are completely dry. Store in clean glass jars with tight-fitting lids. (Careful labelling will prevent any confusion later on.) Choose a cool, dry, dark storage space.

A note on INGREDIENTS

For all my recipes, I prefer to use organic, stone-ground flours. Where possible, I recommend whole wheat or spelt flours. I also recommend using organic milk wherever possible and organically grown produce and meats.

Instead of sugar of any kind, I suggest stevia, which is a natural, plant-derived sweetener without any of the side effects of sugars or artificial sweeteners. It is grown commercially now in Canada, and you can buy it in health food stores. It is extremely sweet, however, so you need only a fraction of the stated amount of sugar in any recipe.

If a recipe calls for olive oil, extra-virgin is best. If a recipe calls for yogurt, the best is a plain organic, naturally thick yogurt without thickeners such as gelatin.

No one can get along in the wild without becoming familiar with its two greatest staples: tea and bannock. These are given special attention here, in keeping with their status. Remember that suggestions for teas occur throughout the field guide, which you can consult for details of the many virtues and uses of individual teas.

Wild Teas

One of the greatest pleasures of harvesting in the wild is the surprising variety of refreshing, fragrant teas you can make with plants you collect. Wild teas can be made from fresh leaves, stems, or flowers as soon as they are picked and cleaned; many can be dried for future use. Be sure you do this properly. If mildew or other forms of decay appear, discard.

Throughout the field guide section of this book, tea-making methods are given for plants that include:

bearberry
blueberry
raspberry
chickweed
wild rose
clover
dandelion
wild sarsaparilla
goldenrod
shrubby cinquefoil
juniper
Labrador tea
pineapple weed
plantain
strawberry
mint

Remember that tea should be carefully made to bring out its full flavour.

Measure cold water and bring to a boil in a saucepan or kettle (avoid aluminum).

Measure the plant parts—bark, stems, leaves, roots, or flowers—drop into the boiling water, and remove from heat; or place in a warmed teapot and pour the water onto them.

To steep, let the container sit for five or ten minutes. The longer you brew it, the stronger the flavour will be. Some teas should be simmered for a few minutes in a saucepan to really bring out their flavour.

Natural teas may need to be strained before serving. Lemon, honey, or sweetener can be added if desired; you may also want to experiment with such spices as nutmeg, cloves, or cinnamon. Many wild teas are also delicious served iced, with a lemon twist or sprig of mint.

Bannock

Bannock was introduced to the New World by Scottish and English sailors who brought bags of flour to sustain them on their voyages. This baked biscuit, a time-honoured outdoor favourite, is easy and versatile. It can be made on the trail or in camp, using a skillet or tin foil over an open fire or baked on hot stones after the fire has died down. It can also be made at home in the oven. Bannock is good warm or cold, spread with butter, peanut butter, or whatever jams and jellies you have made from your foraging. Use your imagination to create sweet or savoury bannocks.

For sweet bannock try any of the following alone or in combinations:

cinnamon
fresh, wild berries (Add at the last minute to prevent crushing. You may want to use extra sweetener if the berries are tart.)
chopped dried fruit
raisins
chocolate chips
nuts
coconut

For savoury bannock try:

onion flakes
grated cheese
herbs, such as wild chives, basil, wild sage, paprika
parsley
sesame or poppy seeds
crumbled bacon

Basic Oven-Baked Bannock

375 mL (3 cups) flour
50 mL (3 tablespoons) baking powder
pinch of salt
5 mL (1 teaspoon) sugar or stevia (optional)
20 mL (1 heaping tablespoon) lard
250 mL (1 cup) cold water

Preheat the oven to 180°C (350°F). In a bowl mix flour, baking powder, salt, and sweetener. Cut lard into the dry mixture. Blend until crumbly. Make a well in the middle of the mix and add water.

Mix in water with a spoon or your fingers. If the dough is too sticky, add a little more flour. Knead lightly, and shape the dough into a flat, round cake. (For small bannocks, divide the mixture into 12 cakes.) Place on a greased baking sheet, and bake for about twenty minutes or until the top is brown. Turn over and bake about 10–15 minutes more, or until the bottom is also brown.

Deep-fried Bannock

Using the basic recipe, make the bannock dough, and divide it into 12 cakes. Poke holes in their centres to make doughnuts. Drop them into hot fat (cooking oil, shortening, or lard), which should be deep enough so that the bannocks float and do not stick to the bottom of the pan. When the bannocks are browned on their undersides, flip them over and cook a few minutes longer to brown the tops. Remove and place on paper towels or newspaper to absorb excess fat; serve hot.

Nutritious Bannock

375 mL (1 1/2 cups) white flour
375 mL (1 1/2 cups) whole wheat flour
50 mL (3 tablespoons) baking powder
90 mL (1/3 cup) skim milk powder
pinch of salt
5 mL (1 teaspoon) sugar or 1/4 teaspoon stevia (optional)
2 eggs
15 mL (1 tablespoon) cooking oil
250 ml (1 cup) cold water

Preheat the oven to 180° C (350°F). In a bowl mix white flour, whole wheat flour, baking powder, milk powder, salt, and sweetener. Lightly beat eggs in a separate bowl. Add cooking oil and water.

Make a well in the dry ingredients and pour egg mixture into it. Mix into a dough, knead a few times, and form into a flat, round cake. Bake on a greased baking sheet for about 20 minutes till golden brown on top, then turn over and bake for 10–15 minutes more to brown the bottom.

Greens

Nutritious wild greens available in the Northeast include the following plants: chickweed, knotweed, clover, lamb's quarters, cow parsnip, dandelion, orach, dock, plantain, fireweed, goldenrod.

The recipes in this section can generally be used for most or all of the greens. Feel free to experiment. Always remember that greens taste best when harvested young, and that some older plants will have to be boiled in one or more changes of water to remove their bitterness.

Baked Eggs in Greens

3 1/2 litres (15 cups) loosely packed, uncooked wild greens
4 eggs (one per person)
pinch of paprika
anchovy, pimento, or sweet red pepper for garnish

Cheese sauce:
15 mL (1 tablespoon) butter
12 mL (3/4 tablespoon) flour
125 mL (1/2 cup) milk
dash Worcestershire sauce
1 mL (1/4 teaspoon) dry mustard
125 mL (1/2 cup) grated cheese (cheddar or similar variety)

Collect enough wild greens, such as dandelions, lamb's quarters, or orach, to fill a 3 1/2 L (3 1/2 quart) pot. Pick the youngest leaves from plants in a dust-free, unsprayed area. Wash them well and chop them up. Put into boiling salted water, and boil 2–3 minutes. Drain off the water. Add more boiling water, boil again, and drain. Add fresh, boiling water a third time and boil the greens until they are soft. (These boilings remove the bitterness that develops in the leaves as the plants mature. In the spring, you only need to boil the greens once.) Drain the greens in a colander and squeeze out the water with a spoon. Set aside, covered, to keep warm.

Melt butter over low heat. Stir in flour and cook for a few minutes to dispel the floury taste. Slowly stir in milk and mix to a smooth paste. Add cheese, stirring constantly; when it's melted, add Worcestershire and mustard. Cover and keep warm over very low heat.

Transfer greens to a buttered baking dish, making four depressions in the greens with the back of a spoon. Break one egg into each. Place under the broiler until eggs are cooked the way you like them—the yolks should remain a little runny. Top eggs with cheese sauce, sprinkle with paprika, and garnish with strips of anchovy. If you're not an anchovy fan, use strips of pimento or circles of sweet red pepper.

This recipe can be adapted for open-fire cooking. Prepare the greens as above, without squeezing out all excess water after boiling. The small amount you leave will evaporate during cooking. Grease a cast iron frying pan and put greens in the bottom. Make the depressions and add eggs. Lay thin slices of cheese over eggs, being careful not to break the yolks. Cover and bake over the coals or in the ashes of your fire.

Fish Baked in Clay with Wild Greens and Herbs

A large fish, pulled from icy waters and pit-baked in locally available clay, makes an incomparable outdoor feast.

1 large fresh-caught fish, cleaned, with gills removed
young wild greens, such as dandelion or fireweed
wild chives or wild sage
lemon (if available)
onion (if available)
clay (common in parts of the region)

Make a good-sized pit in the ground, lining it with stones. Build a substantial fire and let it burn down to embers. Meanwhile, season the cavity of the cleaned fish with herbs, onion, and lemon, and stuff with a few handfuls of tender wild greens. Skewer the opening closed—you want to prevent any clay from getting inside.

The clay should be sticky enough to hold together. If it is dry, work it with water to a dough-like consistency. Mould the clay over fish to enclose it completely, to a thickness of about 5 cm (2 in). Bury coated fish in the hot embers to bake. If necessary, you can rebuild the fire over the fish. A 1 1/2 kg (3 lb) fish will take about 2 hours to bake, depending on the type of fire you build. The clay should be fully hardened when the baking is complete.

Crack the clay to remove the fish. The scales, skin, head, and tail will come off with the clay shell, leaving a moist, delicious feast, and no dishes to wash! If you crack the clay shell and find that the fish is not fully cooked, patch the shell with moist clay and put it back in the fire.

Pan-fried Greens

2 1/2 L (about 10 cups) fresh, young greens (clover leaves, dandelion leaves, fireweed shoots, chickweed, or lamb's quarters)
1 onion, minced
1 tablespoon butter, or olive oil
salt and pepper to taste
lemon juice to taste
fresh-grated nutmeg to taste

Fry onion in butter or oil until translucent. Add greens and stir to coat with butter or oil. Cover and simmer until limp. Serve with salt and pepper, a little lemon juice, and nutmeg.

Rice Pilaf with Greens

750 g (1 1/2 lb) young wild greens
125 mL (1/2 cup) olive oil
250 mL (1 cup) chopped green onions with tops
1 medium onion, chopped
500 mL (2 cups) short-grain rice
60 mL (1/4 cup) lemon juice
30 mL (2 tablespoons) fresh parsley, chopped
30 mL (2 tablespoons) wild chives, chopped
1 small red pepper, diced
salt and pepper

Wash greens well, and drain. If you are using older greens, they will have to be boiled in several changes of water, drained, and squeezed dry. Tear the leaves into pieces and chop the stalks. Heat the oil in a deep frying pan or pot, add the onion, and fry gently until soft. Wash the uncooked rice well, drain, and add it to the onion. Cook for 5 minutes, stirring often. Add all the remaining ingredients except the greens. Bring to a boil, add the greens, and stir well. Cover tightly, reduce heat, and simmer for 15 minutes. Remove from heat, leaving tightly covered. Let the pilaf stand for 5–10 minutes before serving.

Wild Pizza

Pizza dough, for a 30 cm (12 in) (medium) pizza
2 L (8 cups) tightly packed wild greens
2 medium onions
50 mL (3 tablespoons) butter or olive oil
125 mL (1/2 cup) sour cream
1 egg
1 mL (1/4 teaspoon) ground nutmeg
dash ground cloves
dash cayenne
500 mL (2 cups) grated or diced Swiss cheese
60 mL (1/4 cup) grated Parmesan cheese
60 mL (1/4 cup) chopped pimento

Preheat oven to 230°C (450°F). Roll out dough on a floured surface. Fit it into a greased 30 cm (12 in) pizza pan, making a thick rim. Cover and let rise in a warm place for 15 minutes. Bake in the preheated oven for 10 minutes. Remove from the oven, but keep the oven on. Thoroughly wash greens, drain, and chop. Boil in very little water until greens are wilted and the stalks are tender Older, less tender greens may have to be boiled in more than one change of water to remove their bitterness. Squeeze them as dry as possible; you should have about 325 mL (1 1/3 cups).

Sauté onions in butter or olive oil until soft. Add boiled greens and cook until nearly all moisture has evaporated. Cool slightly. In a large bowl, beat sour cream and egg. Add greens, nutmeg, cloves, and Parmesan cheese. Spread evenly on the crust, leaving a border about 1 1/4 cm (1/2 in) wide. Lay remaining Swiss cheese over top. Bake at 230°C (450°F) for 15 minutes or until the cheese bubbles. Garnish with pimento and serve.

Green Cream Soup

250 mL (1 cup) cooked greens
1 onion, raw and cut in half
375 mL (1 1/2 cups) water
2 chicken bouillon cubes
375 mL (1 1/2 cups) milk
salt and pepper to taste
grated nutmeg, pinch
grated allspice, pinch
croutons

In a blender, purée the cooked greens, onion, water, and bouillon cubes. Return to stove and simmer gently for 20–30 minutes. Add milk, nutmeg, and allspice, and heat again. Garnish with croutons and serve. Note that you will need to pick quite a lot of greens to get 250 mL (1 cup) cooked greens—the greens cook down considerably.

Steamed Greens

Choose young, tender greens that won't be bitter. Pick all of one kind, or mix them, but note that some may cook more quickly than others. All of the following can be used: fireweed, dandelion greens, chickweed, lamb's quarters. Wash greens well and put them in a steamer over boiling water, cover, and steam until tender: they will take about 3–5 minutes longer than if boiled. Sprinkle with lemon juice or vinegar, dot with butter or drizzle with olive oil. Salt and pepper to taste, and serve hot.

Buttered Chickweed

500 mL (2 cups) chickweed per person—leaves and tips only
butter
salt
pepper

Thoroughly wash chickweed, then drop it into rapidly boiling, salted water. Cook very briefly, then drain well. Heat a little butter in a frying pan. Add chickweed, turn to coat evenly, and sauté. Add salt and pepper to taste and serve hot, as a vegetable.

Dandelion Green Salad

Mix tender young dandelion leaves, carefully washed, with chopped apple, thinly sliced onion, grated carrots, and toasted walnuts. Serve with an oil and vinegar dressing.

Dandelion Leaf Sandwiches

These sandwiches are similar to the watercress "dainties" served at fancy teas: thick hunks of bread would not be appropriate! The bread crusts should be cut off (freeze them to use later in a bread stuffing).

white or brown bread, thinly sliced, trimmed, and buttered
Worcestershire sauce
young dandelion leaves, washed and very finely chopped

Spread half the bread slices with a few drops of Worcestershire sauce. Sprinkle thickly with chopped dandelion leaves, top with another slice of bread, and cut the sandwiches into four smaller pieces, either triangles or squares.

Wilted Dandelion Greens

5 1/2 L (about 6 quarts) young dandelion greens (or any other greens from the list at the beginning of this section)
6 slices of bacon, fried and crumbled
50 mL (3 tablespoons) cooking oil or bacon fat
60 mL (4 tablespoons) white vinegar
60 mL (4 tablespoons) finely chopped onion
5 mL (1 teaspoon) brown sugar
salt and pepper to taste
3 hard boiled eggs, chopped (optional)

Wash greens well and put them in a large saucepan with just enough water to cover. Bring to a boil, then remove from heat, and set aside for a few minutes. While greens are standing, use an enamel or stainless steel pan to fry bacon until crisp. Remove, drain, and crumble, leaving the fat in the frying pan if you intend to use it in the salad; otherwise, pour off and replace with oil. Stir in the onions and fry until transparent. Add the vinegar and heat briefly, then stir in sugar, salt, and pepper. Cook on low heat for a few minutes. Drain greens well, pressing out excess water, then add to the mixture in the frying pan. Stir and simmer for a few minutes. Garnish with bacon bits and chopped eggs; serve hot.

Sesame and Wild Green Sauté

2 garlic cloves, peeled and crushed
30 mL (2 tablespoons) sesame oil
2 L (8 cups) wild greens (dandelion greens, fireweed, chickweed, knotweed, lamb's quarters, nettle, goldenrod greens), cleaned and coarsely chopped
15 mL (1 tablespoon) tamari or soy sauce
30 mL (2 tablespoons) orange juice
30 mL (2 tablespoons) toasted almonds
toasted sesame seeds for garnish

Heat a wok over medium-high heat and add oil. Sauté garlic for 1 minute, being careful not to burn it (or it becomes bitter). Add greens and sauté for 2–3 minutes or until wilted (sautéeing removes some of the greens' bitterness). Add almonds and orange juice. Season with tamari, cover, and simmer for 2 minutes or until greens are soft. Before serving, garnish with a sprinkling of sesame seeds.

Fiddlehead Sauté

Serve as a vegetable side dish or over toast as a light meal.

450 g (1 pound) fiddleheads, cleaned and dried
butter and/or organic extra-virgin olive oil for sautéeing
125 mL (1/2 cup) onion (wild onions if available), chopped
2 garlic cloves, peeled and crushed
15 mL (1 tablespoon) lemon juice
2 or 3 splashes white wine
5 mL (1 teaspoon) paprika
salt and pepper to taste
sprinkling of sweetener

In a stainless steel frying pan (cast iron will discolour), heat olive oil or butter over medium to high heat. Add onions and garlic and stir-fry for 2 minutes until both are soft but not browned. Add white wine and fiddleheads and continue to steam-fry until fiddleheads are cooked through but not mushy. Add lemon juice, sweetener, salt, and pepper and serve with paprika.

Potato Salad with Fireweed

500 mL (2 cups) fresh, young fireweed leaves or chickweed shoots

1 L (4 cups) potatoes, sliced

bacon, cooked and chopped, for garnish (or use soya bits)

paprika

Dressing:

250 mL (1 cup) sour cream or thick plain yogurt

125 mL (1/2 cup) mayonnaise

60 mL (1/4 cup) fresh dill, chopped; or 5 mL (1 teaspoon) dried

250 mL (1 cup) cucumber, seeded, peeled, and chopped

salt and pepper to taste

Cook potatoes in boiling water until tender but not mushy. Drain well. Do not let them cool.

To make dressing, mix together sour cream or yogurt, mayonnaise, and dill until smooth. Add cucumber, salt, and pepper. Add warm potatoes and greens to the dressing mixture and toss gently, being careful not to break the potatoes, until well combined. Garnish with bacon. Sprinkle with paprika. Refrigerate and serve chilled.

Herbs

Many of the plants you harvest will lend themselves for use as flavourings. You can also use them as aromatics to scent rooms and clothing or even campfires. The most useful cooking herbs include wild sage, wild chives, wild onion, mint, and pineapple weed.

Herb Pudding

This recipe is adapted from an old cookbook of traditional recipes from England and Wales. I bought it from a used bookseller in London many years ago. Usually made in the spring, this was a traditional Easter dish that used as many vegetables as possible—the more included, the more delicate the taste.

475 mL (1 pint) barley or oatmeal, boiled
1 egg, beaten (optional)
cabbage (or any other vegetable from the cabbage family; try curly kale)
onions or leeks
nettles
dandelion leaves
*bistort (*Polygonum bistorta*, an alpine species introduced from Eurasia)*
watercress
sour dock / garden sorrel (Rumex acetosa)
blackcurrant leaves (young)
mint
thyme
salt to taste
pepper to taste
1 tablespoon butter

Wash vegetables and herbs carefully, dry, and chop finely. Mix them with about one pint of previously boiled barley or oatmeal and mis in the beaten egg (optional). Tie mixture into a pudding cloth or pudding basin (which retains more of the juices). Steam for 3–4 hours. Turn cooked pudding onto a plate and season with salt, pepper, and butter. Serve as the vegetable dish of a dinner.

Wild Mint Sauce

500 mL (2 cups) white vinegar
12 sprigs wild mint, washed
2 mL (1/2 teaspoon) white sugar

Using a small enamel or stainless steel pot, bring vinegar to a boil. Bruise mint leaves or chop them finely, and add to vinegar. Boil again, then remove from heat immediately. Add sugar and allow to cool. Pour mixture into a glass bottle. Cork tightly; the flavour will improve with age. Serve with lamb, boiled potatoes, or steamed garden peas.

Cream Cheese with Wild Chives and Herbs

A sandwich spread that goes well with sliced cucumbers, chicken, or ham.

125 g (1 small package) cream cheese
15 mL (1 tablespoon) sour cream or thick plain yogurt
10 mL (2 teaspoons) wild chives, minced
10 mL (2 teaspoons) parsley, minced
2 mL (1/2 teaspoon) chervil, dried; if fresh is available, use 10 mL (2 teaspoons)
2 mL (1/2 teaspoon) tarragon, dried; if fresh is available, use 10 mL (2 teaspoons)
lemon juice, few drops
salt and pepper to taste

Cream together cream cheese, sour cream, and lemon juice, then mix in the other ingredients.

Middle Eastern Pita Bread Salad

2 sandwich-size pita breads (whole wheat if available)
250 mL (1 cup) chickweed leaves and tips, chopped
2 tomatoes, cut into small chunks
1 cucumber, peeled, seeded, and diced
1 sweet green pepper, seeded and diced
3–4 spring onions with tops, chopped
125 mL (1/2 cup) parsley, fresh, chopped
60 mL (1/4 cup) wild mint, fresh, chopped
cherry tomatoes, halved, for garnish
black olives, for garnish

Dressing
125 mL (1/2 cup) freshly squeezed lemon juice
125 mL (1/2 cup) olive oil
1 clove garlic, crushed
5 mL (1 teaspoon) salt
black pepper

Using scissors, cut pita bread into small pieces, and toast lightly under the broiler. Put toasted pita into a bowl with greens, vegetables, and herbs. Shake the dressing ingredients in a covered jar, then pour over the salad and toss well. garnish with cherry tomatoes and black olives.

Middle Eastern Salad Dressing

80 mL (1/3 cup) olive oil
50 mL (3 tablespoons) freshly squeezed lemon juice
5 mL (1 teaspoon) sweetener (honey, sugar, or 1/4 teaspoon stevia)
7 mL (1 1/2 teaspoons) fresh mint leaves, minced
salt and pepper to taste

Combine all ingredients in a bottle. Cover and shake well.

Dandelion Greens Soup with Lentils and Spices

125 mL (1/2 cup) lentils, picked over and rinsed well in cold running water
30 mL (2 tablespoons) organic extra-virgin olive oil
1 medium yellow onion, peeled and chopped
2 cloves garlic, peeled and crushed
2 mL (1/2 teaspoon) dried red chili flakes, or to taste
2.5 cm (1 in) piece of fresh ginger, peeled and finely grated
1 bunch (500 mL [2 cups]) fresh spring dandelion greens, washed well and chopped
30 mL (2 tablespoons) freshly squeezed lemon juice
1 lemon peel (leftover from juice)
2 mL (1/2 teaspoon) ground cumin
salt and pepper to taste

Place lentils and lemon peel in a small-medium cooking pot with 1 1/4 L (5 cups) of water and bring to a boil over high heat. Reduce heat to medium, cover, and simmer until lentils are just tender, about 20–30 minutes.

While lentils are cooking, heat olive oil in a skillet over medium heat. Add onion, garlic, chili flakes, and ginger and cook, stirring often to make sure garlic does not burn (or it will become bitter). Cook until onions are soft, about 10 minutes.

Add onion mixture, dandelion greens, lemon juice, and cumin to lentils. Season to taste with salt and pepper (and more chili flakes if you prefer a spicier dish). Simmer, covered, until dandelions are tender and well-blended with the rest of the soup, about 15 minutes. Serves 4.

Cool Soup

500 mL (2 cups) plain yogurt
250 mL (1 cup) chicken broth, chilled
2 cucumbers
1 bunch red radishes
30 mL (2 tablespoons) white vinegar
30 mL (2 tablespoons) olive oil
15 mL (1 tablespoon) fresh wild mint, minced
salt and white pepper to taste
6 ice cubes
fresh mint leaves for garnish

Peel, seed, and grate cucumbers. Mince enough radishes to make about 60 mL (1/4 cup), leaving a few aside for garnish. Combine with the remaining ingredients, except ice cubes, in a serving bowl and mix well. Chill. Just before serving, add ice cubes and garnish with mint leaves and remaining radishes, thinly sliced. Serves 6.

Mint and Tomato Salad

2 ripe tomatoes, thickly sliced
1 medium-sized sweet onion, thinly sliced
1 lemon, thinly sliced
about 10 fresh mint leaves
olive oil
sweetener (honey, sugar, or stevia), to taste

In a bowl, alternate slices of tomato, lemon, and onion. Chop mint finely and sprinkle on top. Sprinkle with a little sweetener and drizzle with olive oil. (If you are using honey or stevia, blend it into the olive oil.) Let stand for an hour or so before serving.

Chive and Nettle Pesto with Toasted Pumpkin Seeds and Asiago Cheese

500 mL (2 cups) fresh chives
250 mL (1 cup) steamed and drained nettle leaves
175 mL (3/4 cup) pumpkin seeds, hulled
2 medium garlic cloves
175 mL (3/4 cup) grated Asiago cheese (or use Romano as a substitute)
75 mL (5 tablespoons) extra-virgin olive oil
225 g (1/2 pound) butter, softened
225 g (1/2 pound) cream cheese, softened

Toast the pumpkin seeds in a cast iron frying pan over medium heat, shaking frequently so the seeds don't scorch. In a food processor, combine the chives, nettle leaves, and garlic, and chop them to a medium consistency. Move the mixture to a bowl and mix in the Asiago cheese and oil with a wooden spoon to make a paste-like mixture, or pesto.

Blend half of the butter with half of the cream cheese, then add the remaining butter and cream cheese and blend. In the bottom of a bowl that can be used as a mould, spoon one-third of the butter–cream cheese mixture. Spread half of the pesto on top. Add another third of the butter–cream cheese mixture. Add the remaining pesto. Finish off with a final layer of the remaining butter–cream cheese mixture. Refrigerate for several hours until it is hardened. To serve, dip the bowl into a dish of hot water just long enough to loosen the mould and invert it onto a serving plate. Serve with a baguette or toasted bagel chips.

Potato, Chive, Sorrel, and Onion Frittata

A frittata is a flat omelette or sort of crustless quiche (with more substance). It travels very well. Sorrel is not covered in the field guide, but seeds may be purchased for growing in your garden.

15 mL (1 tablespoon) olive oil
(organic extra-virgin if you have it)
30 mL (2 tablespoons) butter
2 medium potatoes, peeled and cubed
15 mL (1 tablespoon) fresh rosemary
(or 5 mL [1 teaspoon] dried)
garden sorrel, wild sorrel, arugula, or spinach
(equivalent to one bunch of spinach)
10 eggs
60 mL (1/4 cup) whipping cream
125 mL (1/2 cup) Gruyère cheese, grated
125 mL (1/2 cup) chives, chopped
chive flowers for garnish
salt
pepper

In a 10-inch skillet, heat oil and 1 tablespoon butter over low-medium heat. Add potatoes and sauté until tender and brown (about 10 minutes). Sprinkle with rosemary and salt and pepper to taste. Set aside and wipe the skillet clean.

To prepare sorrel, pull the stems off backwards and remove any tough centre ribs.

In a stockpot of boiling water, add 5–10 mL (1–2 teaspoons) of salt. Plunge sorrel into the boiling water for 1–2 minutes just until it wilts and has turned grey-green in colour. Drain in a colander and squeeze out any excess liquid (you should have about 350 mL [1 1/2 cups]).

In a large mixing bowl, stir together eggs, whipping cream, Gruyère cheese, and chives. Add potatoes and sorrel.

Preheat broiler. Place the skillet on a burner at medium-low heat and add 15 mL (1 tablespoon) butter. When it has melted, add egg mixture. Cool slowly until the bottom is set, stirring occasionally, for about 10 minutes. (The top should remain slightly liquid.) Place the skillet under the broiler until the top just sets. Cut the frittata into wedges and serve hot, warm, or cold. A fresh tomato salad makes a nice accompaniment.

Serves 4–6.

Petals, Buds, and Roots

Wild plants can be used in surprising ways, many of which are mentioned in the field guide section of this book. The recipes that follow suggest ideas you may not have thought of; they're both adventurous and delicious.

Clover Butter

Pick plenty of fresh clover blooms, removing the petals from their stalks. Mash the petals with sweet butter. Spread this clover butter on very thin slices of fresh bread, scones, or wafers.

Dandelion Coffee

Here is an early recipe for dandelion coffee from a famous guide written for eastern Canadian settlers by Catherine Parr Traill:

"The roots of the Dandelion have been utilized as a substitute for Coffee; in preparing it the root should be washed thoroughly, but the thin brown skin not scraped off, as much of the tonic virtue is contained in this brown covering of the root. This must be cut up into small pieces and dried by degrees in the oven until it becomes dry and crisp enough to grind in the coffee-mill; it is then used in the same way as the Coffee-berry, with the addition of milk and sweetener. A small portion of fresh Coffee would, I think, be an improvement to the beverage, but it is not usually added. Many persons have used this preparation of the Dandelion and greatly approved of it. It is a good tonic and very wholesome." The Canadian Settler's Guide 1855

Goat's Beard Roots

Gather the roots of young goat's beard plants and scrub, scrape, and wash them well. If you prepare the roots ahead of cooking time, place them in water with a little vinegar added to keep them from discolouring. When ready to cook, cut into slices as you would do with carrots. Place in a saucepan of lightly salted, boiling water and simmer until tender; drain. Sprinkle with salt and pepper and serve with butter.

Rose Petal Honey

500 mL (2 cups) honey
250 mL (1 cup) wild rose petals, cleaned

Snip the bitter white "cuticles" from the rose petals with a pair of nail scissors. Bring honey to just before boiling in a small saucepan. Add rose petals. Remove from the heat, let stand for several hours, then reheat to liquefy, and strain into honey pots.

Apple and Rose Petal Pie

pie dough, enough for top only
1 L (4 cups) cooking apples
250 mL (1 cup) water
190 mL (3/4 cup) sugar or honey
4 cloves
rind of half a lemon, washed and grated
a handful of fresh wild rose petals

Wash and dry rose petals. Then cut out white, bitter-tasting "cuticles" with a pair of nail scissors.

Roll pie dough to 1/2 cm (1/4 in) thick and set aside. Preheat the oven to 215°C (425°F). Peel and core apples; boil peelings and cores in 250 mL (1 cup) water. Stew for 1/2 hour, then strain and set the juice aside to cool.

Cut apples into thick slices or chunks and mound them in a pie dish. (Thin slices will let the juice boil out before apples are cooked, and the result will be tough and tasteless.) Add sweetener, rind, and cloves, and pour cooled juice onto the mound. Lay rose petals on top. Wet the edges of the pie dish. Cover pie with the pastry, being careful not to stretch the pastry (or it will pull away from the pie dish in cooking). Trim the edges with a knife. Bake for 15 minutes; reduce oven temperature to 180°C (350°F) and bake for 30 minutes longer or until top is nicely browned.

Rose Petal Vinegar

Collect enough fresh, clean wild rose petals to fill a glass jar, then pour white vinegar over them and cover. "Sun-steep" the mixture by letting it stand in the sunlight for at least two days. Strain out the petals. This fragrant, pink vinegar is good in dressings for both fruit and green salads. Replenish as needed by adding new petals and more vinegar.

Goldenrod Crepes

80 mL (1/3 cup) goldenrod flowers, cleaned and chopped
250 mL (1 cup) all-purpose flour, sifted
1 mL (1/4 teaspoon) salt
3 eggs
375 mL (1 1/2 cups) milk
30 mL (2 tablespoons) butter, melted

In a mixing bowl, blend flour, salt, and goldenrod flowers. Break eggs in the centre of flour mixture and stir flour, a little at a time, into eggs, until half of the flour has been blended in. Add 175 mL (3/4 cup) milk and mix until smooth. Stir in melted butter, then the rest of the milk. Cover and set aside for one hour. Lightly grease a crepe pan or a 15 cm (6 inch) skillet with butter. Pour in 30–50 mL (2–3 tbsp) batter and tilt the pan to make a thin, even coating. Cook until the bottom is lightly browned and the top is dry. Turn over and cook the underside to the same light brown colour. Remove the crepe from the pan and roll up for serving. Repeat the operation, making crepes until the batter is used up.

Berries

Many berries of the Northeast are remarkable for their flavour. The recipes that follow are worth a bit of extra trouble. Berries for most of the recipes have been specified; not all berries are suitable to every use. Still, you may want to experiment, keeping in mind that the recipes are tested carefully for their suitability to the various species.

Berry-Stuffed Baked Apples

6 large apples
190 mL (3/4 cup) wild fruits for stuffing: cranberries with chopped nuts, blueberries with grated orange rind, saskatoons with grated coconut, cranberries with raisins, or saskatoons with granola. (These are suggestions only. Use whatever berries are in season in combination with the cereals or spices you have on hand.)
sweetener, to taste
butter, to taste
cinnamon, nutmeg, or cloves, ground
lemon or orange juice, to taste

Core but do not peel apples. Cut 1″ diameter opening from the "star" ends of apples and set cut piece aside. Set apples on a buttered dish or sheet of tinfoil. Fill apples with a mixture of berries and sweetener. Dot with butter and sprinkle on spices and juice, and replace pieces so that apple is "closed." Bake at 190°C (375°F) for 50–60 minutes. If you're cooking the stuffed apples over an open fire, pull up the sides of the tinfoil sheets into a twist at the top. Then set the packages in the ashes of the fire for about an hour.

Berry Blintzes Stuffed with Cottage Cheese

For the blintzes use:

250 mL (1 cup) all-purpose flour, sifted
1 mL (1/4 teaspoon) salt
3 eggs
375 mL (1 1/2 cups) milk
30 mL (2 tablespoons) melted butter
butter, for greasing skillet

Filling:

500 mL (2 cups) dry cottage cheese
2 egg yolks
60 mL (1/4 cup) sugar or honey (or 1 teaspoon stevia)
60 mL (1/4 cup) sour cream or plain thick yogurt
pinch salt
5 mL (1 teaspoon) vanilla
60 mL (1/4 cup) butter
sour cream or plain thick yogurt for topping
250–500 mL (1–2 cups) freshly picked wild berries: gooseberries, strawberries, raspberries, cloudberries, or blueberries

Mix flour and salt in a bowl. Break eggs in the centre of flour and stir flour, a little at a time, into eggs. When half the flour has been blended in, add half of milk and mix until smooth. Stir in melted butter, then rest of milk. Cover and set aside for half an hour.

Combine all filling ingredients, except for berries and sour cream or yogurt, in another bowl. Lightly grease a crepe pan or a 15 cm (6 in) skillet with butter. Pour in about 30–50 mL (2–3 tablespoons) batter and tilt the pan to make a thin, even coating. Cook until the bottom is lightly browned and the top is dry. Turn this blintz onto a clean cloth, browned side up. Repeat the operation, making crepes until batter is used up. Spoon 30 mL (2 tablespoons) of filling into the centre of each blintz (the brown side). Fold the sides over the filling and turn over; repeat for the rest of the blintzes. Heat more butter in the pan and fry as many blintzes as the pan will hold, turning until lightly browned on all sides. Set aside and keep warm. Fry the rest of the blintzes in the remaining butter.

To serve, top with sour cream (or yogurt) and berries.

Summer Berry and Fruit Bowl

250 mL (1 cup) water
190 mL (3/4 cup) sugar or honey
30 mL (2 tablespoons) lemon juice
15 mL (1 tablespoon) anise seed
1 mL (1/4 teaspoon) salt
fresh fruit, to serve six
(grapes, plums, peaches, pears, pineapple, melon, apples, kiwis, or bananas)
fresh wild berries
(blueberries, raspberries, strawberries, cloudberries, saskatoons, currants, gooseberries)
fresh mint sprigs, for garnish

Combine water, sweetener, lemon juice, anise seed, and salt in a saucepan and cook over medium heat for about 15 minutes, until the mixture thickens slightly into a syrup. Refrigerate until cool.

Prepare fruit. Wash, peel, and remove seeds if necessary; cut into slices, chunks, or wedges. If you are using kiwis, leave unpeeled and set aside with berries. Put remaining fruits into a bowl. Pour syrup over fruit and refrigerate until cooled, giving it an occasional careful stir. When ready to serve the salad, peel and slice kiwis and gently stir into the mixture. Mix in any firm berries (saskatoons, blueberries, currants, gooseberries) but such fragile types as raspberries or cloudberries should be piled on top of the fruit bowl.

Garnish with mint and serve.

Glazed Wild Berry Tart

Here is a recipe that makes the most of wild berries. You can also mix berries with sliced nectarines, peaches, bananas, or other fruit.

750 mL (3 cups) wild berries (blueberries, cloudberries, gooseberries, currants, strawberries, raspberries, or any combination)

Crust

125 mL (1/2 cup) butter
60 mL (1/4 cup) icing sugar, sifted
250 mL (1 cup) pastry flour, sifted

To make the glaze use:

30 mL (2 tablespoons) sugar or honey
15 mL (1 tablespoon) cornstarch
1 mL (1/4 teaspoon) ground mace
125 mL (1/2 cup) orange juice
125 mL (1/2 cup) red currant jelly, melted

Crust:
Preheat oven to 180°C (350°F). Cream butter until soft, then gradually add icing sugar; cream until thoroughly blended. Add flour, a little at a time, to make a soft dough. Pat the dough into a 23 cm (9 in) pie pan, covering the bottom and sides evenly. Bake for 12–15 minutes in the middle of the oven, until golden brown.

Glaze:
Combine sweetener, cornstarch, and mace in the top half of a double boiler. Add orange juice and stir until the mixture is thick and clear. Add melted red currant jelly, stirring until well blended. Cool slightly.

To assemble tart:
Arrange berries in the baked crust. One way to do this (so that everyone gets a taste of all the berries) is to arrange them in concentric circles, using berries you have the least of in the centre, and working outwards. Spoon the glaze evenly over berries, being sure to cover them all. Chill. Serve topped with whipped cream.

Wild Berry Tarts

Pastry

125 mL (1/2 cup) butter

60 mL (1/4 cup) icing sugar, sifted

250 mL (1 cup) pastry flour

Filling

125 g (1 small package) cream cheese, softened

45 mL (3 tablespoons) sour cream

Topping

500 mL (2 cups) fresh wild raspberries, strawberries, blueberries, or cloudberries

60 mL (4 tablespoons) powdered sugar (superfine)

Preheat oven to 180°C (350°F). Soften butter, then cream in icing sugar a little at a time. Add flour, mix together into a dough, and cut into 12 equal pieces. Press lightly into muffin tins, covering their bottoms and sides. Bake for 12–15 minutes; remove from oven and set aside to cool.

Beat cream cheese until fluffy; add the sour cream and beat until smooth. Divide the mixture evenly among the cooled pastry shells. Chill.

Remove chilled shells from the baking pan. Pile berries on the shells and sprinkle them with a little powdered sugar. Add berries and sweetener just before serving so that pastry shells do not become softened by the juice. (I often make the pastry shells and filling before I go out berry picking, then keep them in the refrigerator until serving time.)

Fresh Berry Topping

This is a simple way to use wild berries when they're at their best. The sugar and liqueur release their juices without cooking. The result is a delicious topping to spoon over ice cream, custard, or frozen yogurt.

1 L (4 cups) fresh wild berries (raspberries, strawberries, blueberries, or cloudberries; currants and gooseberries can be used but require extra sweetener)

60 mL (1/4 cup) powdered sugar (superfine)

35 mL (2 1/2 tablespoons) Grand Marnier or Cointreau (optional)

Pick over berries. Wash and drain them if necessary, and dry on a clean cloth. Purée in a food processor or blender, or crush them with a fork until you get a smooth mixture. Sieve purée to remove any unwanted seeds. Add sugar to purée and stir until dissolved. Add liqueur.

Covered and refrigerated, this sauce will keep for several days, or it can be frozen.

Berry–Sour Cream Torte

This recipe is a favourite, given to me years ago by Dufflet Pastries in Toronto. Be sure to make it well ahead of time. It's rich but worth it.

Pastry

250 g (1/2 pound) sweet butter
190 mL (3/4 cup) icing sugar
1 egg, lightly beaten
5 mL (1 teaspoon) baking powder
500 mL (2 cups) all-purpose flour

Filling

500 g (1 pint) sour cream
250 mL (1 cup) icing sugar, sifted
5 mL (1 teaspoon) vanilla

375–500 mL (1 1/2–2 cups) Fresh Berry Topping (see recipe on page 185)

Preheat the oven to 180°C (350°F). Pastry: Cream sugar into butter. Blend in egg and mix well. Sift together baking powder and flour, and add to butter mixture. Divide mixture into three 23 cm (9 in) baking tins. Starting at the centre, pat each piece of dough to the edges of the tin to make very thin, round cookie shapes. Bake for 10–15 minutes, until lightly browned; cool.

Filling: Mix icing sweetener, then vanilla, into sour cream.

Assembly: Spread 1/3 of this filling on bottom cookie, top with second cookie, and spread with second 1/3 of sour cream mixture. Top with third cookie and rest of filling. Keep the sides free of filling. Refrigerate for at least 12 hours. To serve, remove torte from fridge and top with Fresh Berry Topping.

Berry Berry Cake

250 mL (1 cup) unbleached flour
250 mL (1 cup) whole wheat flour
250 mL (1 cup) sugar
20 mL (3 1/2 teaspoons) baking powder
5 mL (1 teaspoon) salt
60 mL (1/4 cup) vegetable oil
250 mL (1 cup) organic soy milk or regular milk
5 mL (1 teaspoon) vanilla
3 free-range eggs
60 mL (1/4 cup) blanched almonds, chopped
250–500 mL (1–2 cups) berries (blueberries or raspberries or a combination works well)

Topping (optional):

500 mL (2 cups) plain natural yogurt
2 mL (1/2 teaspoon) vanilla
15 mL (1 tablespoon) icing sugar (or touch of stevia, to taste)

Preheat oven to 180°C (350°F). In a large mixing bowl, combine flours, sugar, baking powder, salt, vegetable oil, milk, and vanilla. Beat for 30 seconds with an electric mixer.

Add eggs and almonds. Beat for 2 minutes more. Gently fold in berries. Pour into a greased 9 × 13-inch casserole dish. Bake for 35–40 minutes, or until a knife inserted into the centre comes out clean. To make the topping, add vanilla and sugar to yogurt and blend well. Serve warm or at room temperature with topping on the side.

Berries in Whipped Cream

250 mL (1 cup) fresh berries (cloudberries, wild strawberries, or wild raspberries)
250 mL (1 cup) whipping cream
15–30 mL (1–2 tablespoons) confectioner's sugar
2 mL (1/2 teaspoon) vanilla

Whip cream until stiff. Fold in sugar and vanilla; mix well. Clean berries of leaves and twigs, but avoid washing them if you can; if not, drain well and pat dry. Gently stir 3/4 of berries into cream. Spoon into serving dishes and garnish with remaining berries.

As a variation try replacing vanilla with one of the following: almond extract, grated orange or lemon rind, coconut, or liqueur.

Fried Bearberry Mush

Boil cleaned bearberries in water until soft; drain. Sieve to remove seeds. Over medium heat, melt a generous amount of butter in a frying pan. Add bearberries and lightly fry. Serve as a side dish with meat or fowl.

Orange-Cranberry Muffins Stuffed with Cream Cheese

1 large orange
190 mL (3/4 cup) boiling water
60 mL (1/4 cup) vegetable oil
1 egg
500 mL (2 cups) all-purpose flour
10 mL (2 teaspoons) baking powder
5 mL (1 teaspoon) baking soda
2 mL (1/2 teaspoon) salt
220 mL (3/4 cup plus 2 heaping tablespoons) brown sugar, lightly packed (or add stevia to taste)
190 mL (3/4 cup) wild cranberries, cleaned and rinsed
60 mL (1/4 cup) walnut pieces, optional
125 g (1 small package) cream cheese

Preheat oven to 180°C (350°F). Quarter orange and remove seeds. Put orange pieces and boiling water into an electric blender and blend until almost smooth (so that you can still see tiny pieces of peel). Add oil and egg, turning the blender on and off quickly to blend lightly. In a mixing bowl, combine flour, baking powder, baking soda, salt, brown sugar, cranberries, and walnuts. Add blender mixture and stir. Don't overmix or the muffins will be coarse.

Lightly grease a 12-cup muffin pan or use paper liners. Pour a little of the mixture into each cup. Cut cream cheese into 12 chunks, and put one into the middle of each muffin cup. Fill with the rest of the mixture (until two-thirds full). Bake for 30–40 minutes or until golden-brown. The cream cheese eliminates the need to butter these muffins.

Mrs. Niederkorn's Wild Cranberry Pecan Loaf

500 mL (2 cups) all-purpose flour
190 mL (3/4 cup) sugar or honey
2 mL (1/2 teaspoon) baking soda
10 mL (2 teaspoons) baking powder
5 mL (1 teaspoon) salt
60 mL (1/4 cup) shortening
2 eggs, well beaten
190 mL (3/4 cup) orange juice
15 mL (1 tablespoon) grated orange rind
250 mL (1 cup) wild cranberries, cleaned
125 mL (1/2 cup) chopped pecans
15 mL (1 tablespoon) flour

Preheat oven to 180°C (350°F). In a mixing bowl, combine dry ingredients: all-purpose flour, sweetener, baking soda, baking powder, salt. Using knives or a pastry blender, cut shortening in until the mixture is crumbly. In another bowl, mix beaten eggs with orange juice and rind, and pour into flour mixture. Stir just enough to moisten. Coat cranberries and pecans lightly with the 15 mL flour; add them to this mixture.

Use a spatula to pour mixture into a well greased 1.4 L (1 1/2 quart) baking pan. Bake for about 1 hour, until a knife blade inserted into the middle comes out clean. Let cool slightly, then remove from the baking pan. Serve warm or cold. This loaf freezes very well.

Cranberry Vinaigrette

cranberry juice
red wine vinegar
maple syrup or honey, to taste
ground pepper, coarse
garlic, pounded
grapeseed oil, olive oil, or other vegetable oil (or replace with buttermilk)

Mix equal parts cranberry juice and red wine vinegar. Add maple syrup or honey, along with ground pepper and garlic. Mix with oil to taste (a ratio of 1 part vinegar to 2 parts oil works well). You can eliminate the oil for an oil-free dressing, or replace the oil with buttermilk. A salad of mixed greens, toasted walnuts and toasted pumpkin seeds, and crumbled soft goat cheese will take advantage of this vinaigrette's distinctive taste.

Cranberry Ketchup

900 g (2 pounds) cranberries, washed
250 mL (1 cup) water
250 mL (1 cup) vinegar (white or cider)
500 mL (2 cups) dark molasses
2 mL (1/2 teaspoon) cinnamon
2 mL (1/2 teaspoon) ground nutmeg
2 mL (1/2 teaspoon) allspice
6 whole cloves
2 mL (1/2 teaspoon) salt

Boil berries, water, and vinegar in a stainless steel or iron pot until berries soften. Strain the berries, force them through the sieve, then return the berry mush to the pot and add the remaining ingredients.

Red Cabbage with Cranberries

750 g (1 1/2 pounds) red cabbage, cored and thinly sliced
60 mL (1/4 cup) olive oil
250 mL (1 cup) orange juice
250 mL (1 cup) wild cranberries, cleaned
60 mL (4 tablespoons) brown sugar
salt and pepper to taste

Use an enamel or stainless steel frying pan to heat oil. Add cabbage and stir-fry lightly for about 2 minutes. Add orange juice and wild cranberries, then cover and simmer for 3 minutes. Add sweetener, salt, and pepper. Simmer uncovered until the liquid is evaporated (about 3 minutes). Serve hot or cold as a vegetable, or chilled as a salad.

Whole Berry Cranberry Sauce

This sauce can be served with a traditional turkey, and also with duck, cold meats, or cheeses. It keeps well in the refrigerator and can be frozen.

500 mL (2 cups) sugar or honey
625–750 mL (2 1/2–3 cups) water
1 1/2 L (6 cups) fresh or frozen cranberries, cleaned and washed

(Orange or apple juice may be substituted for part of the water.) Heat water in a saucepan and add sweetener, stirring until dissolved. Add berries and simmer for 15–20 minutes, or until the berries are softened and the sauce is thick.

Baked Beans with Juniper Berries

Add juniper berries to canned baked beans or to your favourite recipe for homemade baked beans. Add 4 (blue) juniper berries per serving and simmer with beans. Remove them before serving.

Wild Raspberry Vinegar

1 L (about 1 quart) mild red wine vinegar, white balsamic vinegar, or rice wine vinegar
375 mL (1 1/2 cups) fresh wild raspberries, cleaned, washed, and patted dry

Put berries in a large glass jar. In a saucepan, simmer vinegar for about 5 minutes, then pour over raspberries, filling the jar about 2 1/2 cm (1 in) below the rim. Cap and seal the jar and set it aside in a cool dark place for a few days, until ready to use. After opening you can keep vinegar in the refrigerator for up to 6 months, but check to make sure no mould appears. If it does, discard immediately. You can prolong the life of the vinegar by sieving out the fruit.

Use raspberry vinegar in fruit and vegetable salad dressings, such as one of these vinaigrettes:

Vinaigrette 1

Mix one part vinegar to three parts salad or olive oil, shake with salt, pepper, and a little sweetener to taste.

Vinaigrette 2

135 mL (9 tablespoons) olive oil
45 mL (3 tablespoons) raspberry vinegar
5 mL (1 teaspoon) Dijon mustard
2 mL (1/2 teaspoon) chopped wild chives or minced onion
salt and pepper, to taste

Add mustard, chives or onion, salt, and pepper to the vinegar and mix well. Drizzle in olive oil, stirring with a fork or whisking until blended. One way to use raspberry vinaigrette is over a salad of orange slices, onions sliced paper thin, and toasted pine nuts.

Rosehip Syrup

This syrup keeps almost indefinitely in the fridge. Use it to flavour soda water or a sparkling wine for a rosehip spritzer, or add it to a fruit punch or applesauce.

1 1/2 L (6 cups) rosehips, cleaned (free of dust, stems, and tufts)

750 mL (3 cups) water

750 mL (3 cups) sugar or honey

In a covered, heavy pot, boil hips and water for 15–20 minutes. Strain through a jelly bag or a piece of clean cloth to remove seeds and sediment. Return this juice to a cleaned pot. Add sweetener and boil for about 5 minutes, until thickened.

Pour cooled syrup into a bottle (it doesn't need to be sterilized and sealed) and store in the refrigerator.

Open-faced Wild Strawberry Sandwiches

250 mL (1 cup) fresh wild strawberries, well cleaned

1 orange peel, grated

20 mL (1 1/2 tablespoons) freshly squeezed orange juice

125 g (4 ounces) cream cheese

4–6 slices of brown bread, with crusts removed

freshly ground pepper (optional)

Wash strawberries (if needed), dry on paper towel or cloth, and set aside. In a small bowl, mix together cream cheese, orange juice, and grated peel. Whip this cheese mixture with a fork until fluffy and spread thinly onto bread slices. Top with plenty of berries and sprinkle with very little ground pepper. (If you've never tried this unusual combination you're in for a pleasant surprise.) Serve right away.

Crowberry Lemonade

Crowberries are not as popular as other berries for puddings or condiments but they are juicy, and quite usable as a beverage, especially if you remove the small seeds.

1 L (4 cups) crowberries, cleaned

Juice of 1 lemon

Sweetener, to taste

Fresh mint, sprig

In a medium-sized pot, bring crowberries to a boil in 2–3 times as much water. Reduce heat and simmer for a few minutes until berries are soft. Remove from heat and strain out seeds and skins through a piece of light cotton or cheesecloth. Add sweetener and lemon juice to remaining liquid. Chill and serve with mint and a slice of lemon or orange over ice.

Historical Sources

A number of sources, both contemporary and historical, were consulted in the preparation of this book; some of them are quoted. The extracts are taken from the writings of explorers, missionaries, and scientists who came to the Northeast as keen observers and recorded their experiences to the great enrichment of future generations. For those who may be curious to know more about them and the plants they wrote about, this listing of sources goes beyond bibliographic information to take a look at the writers themselves.

Beavan, Emily (1818–1897)

Born in Ireland, Beavan arrived in New Brunswick in 1836 and became a teacher. Her father was a sea captain who visited Saint John, NB, and Yarmouth, NS; her husband was a physician who practised in New Brunswick. She was a poet and the author of tales about New Brunswick as well as *Sketches and Tales Illustrative of Life in the Backwoods of New Brunswick, North America* (George Routledge of London 1845). *Backwoods* was based on seven years of residence in New Brunswick and gave her observations about settlers' lives and conditions before the family moved back to England. The book is available online.

Biard, Pierre (c. 1565/7–1622)

Biard headed the first Jesuit mission to Canada, arriving into Nova Scotia. He was one of the founders (1613) of the French settlement at Bar Harbor in what is now Maine. His descriptions of the local Algonquians (especially the Mi'kmaq of Nova Scotia) were compiled into the Jesuit *Relations*, providing a contemporary record of Mi'kmaq culture and of French-Aboriginal relations in the Northeast.

Jesuit Relations II (1612–14), *III* (1611–16), translated into English by Reuben Gold Thwaites; *Relation de La Nouvelle France* (1616).

Bonnycastle, Sir Richard Henry (1791–1847)

An English soldier who served in Canada during the war of 1812, Bonnycastle wrote four books based on his travels throughout the colonies, including *Newfoundland in 1842: A Sequel to "The Canadas in 1841"* (2 volumes. London 1842).

Cartier, Jacques (1491?–1557)

Born in Brittany, Cartier sailed along the coast of Newfoundland, calling it "the land that God gave to Cain." He "discovered" Prince Edward Island on the same journey. His travels established the basis for France's claim to territory in what became Canada. Cartier often confused European and Canadian species of flora and fauna. Like other early explorers, his interests in new plants were economic—he listed plants with edible fruit as well as woods, oils, and plants used by the Indigenous Peoples he met or heard about.

"A Shorte and Briefe Narration" (Cartier's Second Voyage, 1535–1536; first published in Paris in 1545) from *Early English and French Voyages, chiefly from Hakluyt, 1534–1608* (Henry S. Burrage, ed. New York 1906).

Champlain, Samuel de (1567–1635)

Champlain made his first voyage to Canada in 1603 as a passenger. In 1604, he arrived in Acadia where he spent three years exploring and mapping the coast and became known as the Father of New France.

The Voyages and Explorations of Samuel de Champlain, 1604–1616, Narrated by Himself translated by Annie Nettleson Bourne, together with 'The Voyage of 1603,' reprinted from 'Purchas His Pilgrimes', Edward Gaylord Bourne, ed. (Two volumes. New York 1922. First published Paris 1632). *Voyages of Samuel de Champlain 1604–1618.* W.L. Grant, ed. In *Original Narratives of Early American History.* New York 1907.

Charlevoix, Pierre François Xavier de (1682–1761)

A Jesuit who arrived in Canada in 1705, Father Charlevoix travelled down the Mississippi as far as New Orleans. His history of New France includes botanical notes and references to the sunflower, ginseng, poison ivy, and corn.

Journal of a Voyage to North America (London 1761. First published France 1744).

Cook, James (1728–1779)

A Yorkshire-born seaman in the Royal Navy, Cook spent five years charting the Newfoundland coast. His map-making skills led to his appointment as surveyor of Newfoundland and to his first expedition command in 1768.

The Voyage of the 'Resolution' and 'Discovery' 1776–1780. J. C. Beaglehole, ed. (2 volumes. Cambridge: Published for the Hakluyt Society, Cambridge 1967).

Denys, Nicolas (1598–1688)

By 1632, Denys was in charge of an expedition to Acadia to take possession of the country and establish a colony. In Nova Scotia, he established the first in a series of enterprises to do with the fishing, lumber, and fur trades; these extended through Cape Breton, New Brunswick, and the Gaspé. Cape Breton established the Nicolas Denys Museum as a tribute to his life and work as Cape Breton's first permanent settler.

The Description and Natural History of the Coasts of North America. (Translated and edited by William F. Ganong. Toronto: The Champlain Society, 1908. First published in France 1672 as *Historique des l'Amérique*).

Dièreville, Sieur de (birth and death dates unknown)

In 1699, he travelled to Acadie where he gathered plants and made observations about the Aboriginal People and the French. About twenty-five specimens of Acadian plants are labelled with his name in the herbarium of the Musée de l'histoire naturelle in Paris. The species *Diervilla* (honeysuckle) was named after him. Upon his return to France in 1700, he became a surgeon, perhaps incorporating the botanical knowledge he gained in Acadie into his practice.

Relation of the Voyage to Port Royal in Acadia or New France. Translated by Mrs. Clarence Webster (John Clarence Webster, ed. Toronto: The Champlain Society, 1933. First published in France 1708).

Ewing, Juliana Horatia (Gatty) (1841–1885)

Yorkshire-born, Ewing was an author of children's literature. She spent two years in Fredericton, New Brunswick, with her husband, Major Alexander Ewing, who was stationed there as an army officer. A digital copy of her book is available online, including illustrations of primrose and willow made while she was in Fredericton. It includes letters and sketches she sent home to her family in England, to whom she also sent gifts of beadwork and basketry she had bought from her friend, Peter Polchies, a Maliseet. She kept a diary and a "commonplace book" of her stay in Fredericton.

Leaves from Juliana Horatia Ewing's 'Canada Home' (Gathered by Elizabeth S. Tucker. Together with Facsimiles of Eight Water-Color Drawings by Mrs. Ewing's Own Hand. Boston 1896).

Fisher, Peter (1782–1848)

A lumber merchant, Fisher was New Brunswick's first historian. He published anonymously, however, as "An Inhabitant of the Province," and wrote these inspiring words: "Whatever concerns my country, interests me / I follow nature, with truth my guide."

Sketches of New Brunswick containing an account of the first settlement of the province with a brief description of the country, climate, productions, inhabitants, government, rivers, towns, settlements, public institutions, trade, revenue, population, & c. (1825).

Gaultier, Jean-François (1708–56)

Appointed King's physician, in 1741 Gaultier came to Canada, where he was a physician in Quebec. In 1747, he requested that all the forts in New France assist in the collection and cataloguing of specimens that he then sent to Paris. The genus *Gaultheria* (wintergreen) was named after him. In 1951, his four-hundred-page manuscript on Canadian plants was found.

Goldie, John (1793–1886)

He studied at Glasgow's Botanic Gardens. From 1817–1819, Goldie visited North America, but his collections were lost through various misfortunes. A fern, *Aspidium goldianum*, was first identified by, and later named after, him.

Head, Sir George (1782–1855)

In the British Army, Head was stationed at Halifax. His book of Canadian reminiscences is *Forest Scenes and Incidents in the Wilds of North America* (2nd edition. London 1838. First published London 1829).

Hébert, Louis (1575?–1627)

Canada's first colonist, Hébert was an apothecary and later a farmer who grew grain and vegetables and owned an apple orchard in which he grew trees from Normandy.

Lahontan, Louis-Armand de Lom d'Arce, Baron de (1666–1715?)

Lahontan came to Canada in 1683 as an officer. He settled in Newfoundland until he was forced to leave by the governor for insubordination. His books have been popular and have been republished many times, sometimes in translation.

New Voyages to North-America. Reprinted from the English edition of 1703 (Reuben Gold Thwaites, ed. 2 volumes. Chicago 1905. First published The Hague 1703).

LeClerq, Chrestien (1641–?)

A Recollet missionary who was critical of the Jesuits, LeClerq travelled to the Miramichi River, leaving the Mi'kmaq by 1686. His system of hieroglyphics is still the basis of that used by the Mi'kmaq today.

New Relations of Gaspesia. Translated and edited with a reprint of the original. (William F. Ganong. The Champlain Society, volume 5. Toronto 1910. First published France 1691). Reprinted in bilingual French-English edition by Invisible Books online.

Lescarbot, Marc (c. 1570–1642)

A French lawyer and author who viewed First Nations positively and critiqued French society of the time, Lescarbot accompanied his client, Seigneur de Poutrincourt, to what is now Nova

Scotia. Returning a year later, he published several accounts of "New France" including, *Histoire de la Nouvelle France* (Paris: Jean Milot 1609. Followed by two editions in 1612–1613 and 1617–18).

Maillard, Abbé Antoine Simon (1710–1762)

Born in Chartres, France, he was ordained circa 1734. He arrived in Louisbourg in 1735 and went to his mission to live with the Mi'kmaqs. He was a linguist who spoke fluent Mi'kmaq, and in 1738, he worked out a system of Mi'kmaq hieroglyphics. In 1745 with the fall of Louisbourg, he was expelled along with about two thousand other French inhabitants, returning to later become a paid Indian agent for the British and a supporter of the Mi'kmaq.

Rand, Rev. Silas Tertius (1810–1889)

Born in Cornwallis, Nova Scotia, he was a missionary to the Mi'kmaq in Nova Scotia for forty years. Fluent in Mi'kmaq and Maliseet as well as other languages, Rand was the author of a dictionary and other works on Mi'kmaq language and culture, including *Legends of the Micmacs*. 2 volumes. (New York and London: Longmans, Green and Co. 1894).

Pacifique de Valigny, Father (1863–1943)

A French Capuchin missionary, Father Pacifique began his mission with the Mi'kmaq in 1894 in Restigouche, Quebec. He published articles in the Mi'kmaq language and in scientific journals, translated prayers, psalms, and catechisms into the Mi'kmaq language, produced a Mi'kmaq grammar, and developed the Pacifique Orthography.

The Micmac Grammar of Father Pacifique (translated and retranscribed by John Hewson and Bernard Francis. Winnipeg: Algonquian and Iroquoian Linguistics 1990).

Sarrazin, Michel (1659–1734)

Trained in medicine, Sarrazin arrived in Canada in 1685. He returned to Europe where he furthered his studies in medicine and also botany at the Jardin Royale des Plantes. Back in Canada again, he made contributions to medicine, botany, and zoology, including sending plants and seeds back to the Jardin Royale. The pitcher plant, *Sarracenia purpurea*, was named after him.

Speck, Frank G. (1881–1950)

An anthropologist who published many articles and books on the ethnography of the region, including aboriginal uses of plants.

Verrazzano, Giovanni da (c. 1485–c. 1528)

The first European to sail the coast of North America, Verrazzano's six-month voyage took him from Florida to Newfoundland. His letter is the earliest description of the North Atlantic coast and the earliest account of Indigenous People north of Mexico.

The Relation of John Verarzanus, a Florentine, of the lands by him discovered... 1524 from Divers Voyages Touching The Discovery of America (J. W. Jones, ed. Published by the Hakluyt Society, 1st series volume 7. London 1850. First published London 1582).

Index

D

E

F

G

M

N

O

P

R

S

Image Credits

All images and prints courtesy of the author, with the following exceptions:

illustrations in part one: Harter, Jim. *Plants: 2400 Copyright-Free Illustrations of Flowers, Trees, Fruits and Vegetables*. Dover: Mineola, NY, 1988.

page 134 images: Dr. Rob Ireland

The author is available for lectures and workshops on ethnobotany, medicinal plants, wild foods, nature-based gardening, and nature printing. Please contact the author if you would like to be notified of upcoming events.

For research and classroom use, a CD of colour photographs is available for purchase (with partial proceeds supporting an award recognizing the advancement of traditional plant knowledge in the Maritimes).

Contact:
Marilyn Walker, PhD
Associate Professor
Department of Anthropology
Mount Allison University
Sackville, NB E4L 1B3
mwalker@mta.ca